A Very Small Farm

A Very Small Farm

by

William Paul Winchester

Illustrations by Carol Stanton

COUNCIL OAK BOOKS
TULSA, OKLAHOMA

Council Oak Publishing Company, Inc.
Tulsa, OK 74120

First edition
00 99 98 97 96 8 7 6 5 4 3 2 1

ISBN #1-57178-021-1

Book and jacket design and illustration by Carol Stanton

If one advances confidently in the direction of his dreams, and endeavors to live the life which he has imagined, he will meet with a success unexpected in common hours.

—Henry David Thoreau
Walden

TABLE OF CONTENTS

There is an air of
permanence and stability…

There is an air of permanence and stability about Southwind Farm. The house and barn, of stucco with hip roofs, have a solid look about them, vaguely Cape Dutch, or maybe Provençial, now that the poplars have grown taller and I've painted the house and outbuildings terra-cotta with white trim and blue shutters.

To the right of the gravel drive winding up to my house is a stand of beehives, a small greenhouse (made from someone's discarded storm windows), root cellar, vineyard, and winter garden. To the left is my orchard (the tree nearest an Arkansas Black, its branches heavy with apples) and beyond that a quarter acre of garden. Behind the house is the poultry yard, a peach orchard, a small wood I've planted, and pasture for Sophia and her calf.

Already bred, Isabel will freshen this fall to become my milking cow, Sophia staying on for sentimental reasons. I've persuaded myself the manure she produces for the garden will earn her keep. Isabel's milk and the registered calf she bears each year will more than earn hers.

The brown eggs from my Buff Orpingtons are also much in demand, and the cost of a household flock is chicken feed. But what really makes it economical are those eggs for my table and the dozen hens I put up for winter meat, after they have enjoyed their perfect summer. The sound of their clucking under my window and the rooster crowing at daybreak I count as clear profit.

The produce from my garden is so abundant it's been five months since I was last at the grocery store, and then for so few items I can still list them from memory: tea, salt, vanilla, cocoa, yeast, spices, spaghetti, and olive oil ... my tastes inclining to the Mediterranean.

I could have got by with less. By the end of harvest my pantry is stocked for the winter with jars of canned asparagus, peas, carrots, zucchini, green beans, sweet corn, tomatoes, peaches, grapes, pears, plums, applesauce, cider ... with sacks and canisters of dried corn, grain amaranth, sesame seeds, peas, okra, beans (colorful as jewels), fruit leather, figs, peaches, and plums. In the freezer are mulberries and blackberries (both gathered wild), green peppers, cantaloupe, watermelon juice, eggs (first frozen in ice cube trays), hens, pecans, milk, and butter ... and in the refrigerator are beets, garlic, radishes, and Swiss cheeses in red wax. In the root cellar (some in trays of moist sand) are potatoes, onions, winter squash, turnips, apples, and pears. All this abundance produces a sense of infinite well-being, but none of it, except for a little from the sale of milk and eggs, is in the form of cash.

For that I look to my bees. Southwind Farm is an apiary, producing from fifty-odd hives something over a ton of honey and honeycomb. This I sell from my house. "Pure Wildflower Honey," the sign reads ... and well before the next season, "Honey Sold Out, More In July." If the immediate area would support more hives, or if I cared to truck my bees farther afield, I might have been entirely self-sufficient from my farming.

As it is I could manage, at least through hard times. Even if that is an illusion, it is not an abstract one; the bees have laid up ample stores for themselves, the pantry and cellar are filled, the winter wood is stacked, the hens laying, the cow fresh, and the fallow garden ready for spring planting. The basic economy of my peasant farm works so well I sometimes think of it as a little universe in which everything is fixed in orbit as if for an eternity.

Still, it is not a perfect economy, my farm. No economy ever is, and I have to make allowances. The Jersey cow, for example, is not as cost-effective as the fifty beehives, my chief source of farm income. But there are those Swiss cheeses to consider, and I do enjoy my cow. In this respect she is like my dog, Berenice, or the cats. And things will go wrong, as when a rainy spring reduces the nectar flow and there is less honey to sell. Even in a good year there is never enough cash income.

To make up the difference I recently took on the job of doing all the brush mowing for a neighboring ranch, buying the tractor and implements with money I had earned as a substitute teacher and caretaker for a small church. It's closer to home than either of those jobs and—with the swooping barn swallows to keep me company through the afternoon—I enjoy mowing.

Any more than this seasonal work and I would cease to be a farmer, something I very much want to remain—a *small* farmer, with garden and orchard and vineyard and bees and poultry and cow, getting my living in the pleasantest way imaginable.

Even the supposed hardships—the work, the solitary life, the staying put, the doing for oneself, the frugal existence—are not as they appear.

The work of a small farm is not so much hard labor as it is a matter of keeping up with things. The okra that isn't picked today will be tough and stringy tomorrow, and the cow has to be milked every morning at seven and every afternoon at five.

For company I have friends and neighbors, my animals and the life of the field. Sometimes solitude is the best of company. Other times I have wished for a farm community, but not enough to have spent my life looking.

Travel I prefer to do in my own way, in the books of Conrad and H. M. Tomlinson and 'Shalimar.' The life at sea they describe is so familiar, so like my own that I sometimes think of Southwind Farm as a small ship outfitted and provisioned for a long voyage, which it is. Besides, I can't be away without feeling I've missed something at home.

As for doing for oneself—well, that's the whole point. On a small farm you expect to live by your wits, teaching yourself to do whatever has to be done. From every quarter we are insistently reminded how incapable we are, how ignorant and unskilled, a hundred professions shouldering us aside to do what we are perfectly able to do for ourselves, leaving us with a vague sense of inadequacy. It almost amounts to a conspiracy, this effort to deprive us of the pleasure in accomplishment.

As it is, I couldn't afford the services of that economy. And there isn't much it can do for me, life on a small farm having changed little. Frugal as the existence sometimes is, it is a marvelous economy. No other enterprise is, in a real and tangible way, as productive as a small farm. Some of that okra I'll have for dinner, sliced and rolled in cornmeal and fried in a hot skillet. The rest I'll dry in the sun for gumbo later this winter. And I've a customer stopping by for two gallons of that Jersey milk, the money going mostly for hay and feed, my profit coming in the form of cheeses.

The sparse economy of Southwind Farm is so inextricably linked with my reasons for being here in the first place that I can't decide which is means and which is ends. Am I sleeping on my screened-in porch in summer and reading before my hearth in winter out of necessity or preference? Doing without air-conditioning and modern heating is a small price to pay for the companionship of a night-singing chuck-will's-widow and a wood fire.

Having given up nothing without getting more in return, I find that the frugality doesn't really matter. It was for the most self-indulgent reasons I came to the farm—to be happy. Everyone knows that happiness is largely a matter of being content. But content with what? The answer to that requires an act of will, which in my case took the shape of a small farm.

To live in the country in a house I built for myself, with meaningful work and a margin of leisure, free to create a little universe of my own making—*this* was my idea of happiness.

Even if something should come along to snatch the farm away, I would go on living as the experience has taught me—deliberately, taking things as they come, in control as much as possible of my own destiny, and farming as I could. In a pot, if it came to that, in a south window.

The life of the farm ...

The life of the farm fascinates me. I could have wished for nothing better. And yet it is almost by accident that I am here.

I grew up in the country, on land belonging to my grandparents, but not on a farm. Except for distant relatives, sepia photographs in old albums, there were no farmers in my family. My father was a teacher, commuting to his university job in town. My mother was a housewife, a word that sounds quaint now. It was she who introduced the idea of a small farm, I think, reading to my younger brother and me every afternoon—those first books supplying the details of an idyllic landscape, the familiar orchard and meadow and barnyard, its animal inhabitants straight out of fable. And all of it, as I discovered from experience, in its way true.

Just as instincts are. If there is an inherent motive for self-preservation, one specific enough to include the hunting instinct you sometimes hear about, then there must be a farming instinct—a latent talent for growing food and tending animals.

A collared lizard lived with us for many years, in a traveling cage when we went on vacation, dining on grasshoppers I caught for him at highway rest stops. An exotic seedling given me about the same time, a reward for letting the doctor look in my aching ear, survived to become a tree, a palm imposing enough to suggest the tropics.

—Which is where I had my first real look at a small farm, by leaning out a hotel window in the West Indies. It was a little hotel, ours a room on the second floor. The windowsill had been so thickly painted it had the smoothness and coolness of porcelain. And from there I could look down on a primitive farm.

On the slope below, screened from the hotel terrace and from everyone (I imagined) but me, a man was clearing a piece of land the size of a tennis court. With a machete he hacked the spindly stalks off at ground level, saving the stouter ones after he had stripped away the foliage. Some he had already used in the pen for his poultry, weaving them basket fashion on stakes driven in the ground. I had never seen a wattle fence, and I'm not likely to see another. Everything about the man's small farm displayed the same resourcefulness, and nothing impresses a child of a certain age quite as much. The pigsty he had made of heavier stuff, mostly driftwood picked up along the beach, salt-stained and still encrusted with rock barnacles. Even his shanty was all salvaged materials, something Robinson Crusoe might have built.

When he had cleared enough ground to suit him, squared it off and spaded it up, he planted his first crop—all in the space of an afternoon and while I watched. I knew from my experience with gourds and sunflowers that he would have to wait for his seeds to grow, and that some of them would never come up. And I also knew from the death of my one-eyed bantam rooster that chickens were a chancy business. But I never doubted he would succeed in the end. He was probably a squatter, and in any case a bulldozer will have been there by now. At the time though it seemed to me his farm would last forever. Some dreams do outlive the dreamer.

It was two years before I saw another small farm, this one in Spain. The scoured look of Andalusia and its sharp wind could hardly have made a more dramatic contrast with the tropics, and yet its peasant farms were familiar from that earlier glimpse on St. Vincent. The landscape of Spain reminded me more of what I had known, and yet the peasant farms were different from anything I had seen in Oklahoma.

My father had run across an ad in the newspaper for a nine-day tour that coincided with spring break. Madrid was still locked in winter, but in the South the pale pink blossoms of almond and apricot were everywhere. And so, if I looked closely, were the small farms—their vineyards and twisted olives and gardens enclosed by low walls of loosely laid rock, the same rock that anchored the sheet metal roofs and sometimes materialized into the lean black poultry of the region or a goat.

On that sweep of arid plain the peasant farms had an aloof and defiant look that appealed to a boy's sense of independence. At night their lights twinkling in the distance reminded me of a lighthouse. (Even now, walking home after dark, crossing the meadow from a neighbor's house, the pale light from my window never fails to evoke the same impression.)

Half fable, half peasant farm—Southwind is plainly the product of all those early experiences. Even though my livelihood depends on it, the farm is in some way child's play. That black farmer on St. Vincent remains for me the very exemplar of industry and resourcefulness. And Southwind Farm is, with its cluster of outbuildings and pens and garden plots, like those peasant establishments on the plains of Andalusia, a refuge and a citadel.

Until nearly through college I never gave much thought to what I would do for a living, and none at all to farming. With a meadow to play in and long summers, I had got used to doing pretty much what I wanted, carving hieroglyphics in the weathered sandstone outcroppings or just watching the cumulus clouds trail their shadows across the prairie landscape. The farms pointed out to me from the car window, ranches and wheat fields, didn't interest me in the least. Even if the university had offered agronomy or horticulture, it is doubtful I would have gone that route.

As one by one my friends settled on a profession they became preoccupied with learning their parts, and for me it was a lonely time. There had never been many jobs for botany majors and none was likely to come along. I wasted a lot of time thinking about what I might have done. But envying the exploits of Charles ("Chinese") Wilson, a nineteenth century plant explorer, plainly wasn't getting me anywhere.

The best that can be said for this difficult time is that it forced me to think seriously about occupations and existences, how people spent their lives at work and leisure. I disliked interference, and I was impatient with boredom and with any activity that didn't yield tangible results. And there were, I realized, a great many things I didn't want to do and wouldn't do—hardly a promising attitude for someone who had to find a job.

In the end I'm not sure that what I decided on is even a profession. Unlike agribusiness, a small farm is not a way of making a living so much as it is a way of life. And, however he comes to it—or she, for the first farmers were women—the small farmer always arrives at his destination by a circuitous route and sometimes an anxious one.

What I wanted then was a way of life, and in that of a small farmer I have not been disappointed.

TWENTY ACRES

Twenty acres was more land than I wanted, but less than the present owner had any use for. It didn't fit, lying across the road from the rest of the ranch—

which was to the east, open land with a broad horizon. Whatever else my small farm lacked, it would have a prospect and a sunrise.

The ground was still faintly furrowed from years ago when someone had tried wheat and before that field corn. Since then it had been planted in bermuda and sweet clover in a half-hearted effort to make a pasture. Some of the bermuda remained, mostly where I wouldn't want it, in the only likely spot for an orchard and garden, the site of the old barnyard.

Here there had once been a farm of which little remained but scattered bricks, a root cellar, and hundreds of daylilies. This was the Old Dutchman's Place. I would have preferred building where no one had lived before, which shouldn't have been hard. Little more than a hundred years ago these were the hunting grounds for the tribes of the plain. In the end I was grateful to him for having made a start.

The site was dominated by a huge mulberry, its trunk more than a dozen feet in circumference and well buttressed. As with old mulberries, there were several dead limbs in the center of the crown that would have to come out before I could build anything beneath, limbs too massive for me to handle. The black mulberry isn't native to North America, and this one had undoubtedly been planted early in the century by the old Dutchman himself—whom I was by now beginning to think of as a collaborator.

He must have planted the other mulberry as well, at one time reduced to a gnarled stump from which sprang a successor, itself now a wide spreading tree. And the two large redcedars out front would also be his—and the line crew's, who pruned them back each winter. The stand of a dozen slender ash and hackberries was on the other hand "volunteer." Encouraged by this little upland wood I later planted a larger one.

The old Dutchman's root cellar, full of debris, was otherwise sound. Most of these early cellars were roofed by arches of mortared sandstone laid in vertical wedges, elegant as a Roman ruin but nothing you'd want to stand under. His was a slab of poured concrete, still free of cracks, which I now use as a threshing floor. I looked everywhere for a date—it's traditional. Even his hens had left their tracks in the concrete rim of his well. Finding none, I concluded that my Dutchman, at least when it came to wet cement, was unsentimental. But that was before I got to know him.

He's always turning up, his tools, whetstones, bottles, broken china, hardware, pieces of farm implement, once a yoke with brassy bosses and big enough for an ox, a long-handled pump of cast iron, mother of pearl buttons—enough buttons to fill a box. I can't imagine how he lost so many and still had both hands free for his plow, whose moldboard I unearthed this past spring, my garden a veritable Lost Dutchman's Mine.

Recently some boards salvaged from his house were pointed out to me, now in the floor of a neighbor's shed. The wood had aged to the color of dark amber, except where a mottling of paint still adhered—canary yellow. None of your traditional farm house white; he liked color. From all the lineament and snuff bottles I unearthed, I had until then taken him for a fairly stolid fellow. Obviously I misjudged him.

Others, too, have come and gone. The Old Claremore Road just nicks the southwest corner of the property. Near sunset on a clear evening or after a light drifting snow, the track is sharply defined. In muddy weather it must have been a hard pull for the wagon teams, up the long hill from the Verdigris River. I once met a man who as a boy watered the horses and mules from a well just over the rise, for a penny. (Years later he drilled water wells for the British Army in North Africa.) On the day I went to look at the land, though, the Old Claremore Road was nothing more than a puzzling feature of the topography.

Like that shallow depression lying just inside the south fence line. What I took for a buffalo wallow had been the site of a plane crash in the early twenties. The engine still lies buried there—worth digging for, they say, but I don't intend to. Chorus frogs breed there every spring when for a few weeks it's a small marsh.

Nothing is as it appears, there is always a story. A few yards east of where the plane went down and older even than the Old Claremore Road is an ant hill. It has been there for as long as anyone can remember, surviving the drought years of the thirties and repeated plowing. A solitary clump of butterfly weed, in bloom a flamboyant orange, marks the spot. A bare and sandy wash, it's hardly the place to establish a city. But these harvester ants belong to an industrious race. Long ago they built a road, eighty yards and for most of its length nearly an inch wide, to more luxuriant lands by the side of a pond. After I came and established my farm they extended their road another sixty yards.

We are neighbors now. On July evenings I've watched caravans of harvester ants, each with a cheat-grass seed gripped in her mandibles, returning along this road, their passing feet leaving delicately scored tracks in the dust. And in mid-August I've noticed winged males and queens gathering at the entrance of the ant hill in preparation for mating flight, this being the time of year when new colonies are established. The fate of these colonists, however, remains a mystery. Except for this hill, half a century old at least, there are no other harvester ants on my twenty acres.

Like the ants, I would be getting my living from only a part of the land, the most fertile. The rest I would pasture lightly if at all. Despite having been overgrazed, there was nothing wrong with the soil an application of rock phosphate wouldn't put right. Even in this there was an element of the unexpected. Lightning bugs and glow worms, their bioluminescence dependent on phosphorus, were drawn to my twenty acres in such numbers that walking out on a still summer evening is like passing through the center of a meteor shower.

To anyone who lives in a meadow with a pond they are familiar company, these drifting sparks, and strangely reassuring. Even in the darting flash there is something deliberate, measured—the flashes coming in a definite sequence, eight or ten to the minute, with first a shorter then a longer interval between. Early in the evening or on warmer nights the tempo quickens to perhaps fifteen or sixteen flashes to the minute. I don't know why it is, though, that on some nights the lightning bug, instead of skimming the pond or hovering above the grass, swings so high its flash is lost against the stars.

One by one the twenty acres is yielding up its secrets, to which there will never be an end. Some belong to the past and some to the future. But on the day I went walking on the land there was just the present moment.

I stood then on what was for me an empty continent. These were the southern plains and this the Cherokee prairie, a wedge of gently rolling grassland stretching up into Kansas and down to the sands of the Arkansas River. To the west lie the Osage Hills, pink in the distance, and to the east, just over the horizon, the Ozark Mountains. These meadows and pastures—wherever you find big bluestem, little bluestem, switchgrass, Indian grass—are surviving fragments of the great tallgrass prairie. There are woods of oak and hackberry on the hills and sycamore, elm, and persimmon along the draws, but the prairie is dominant, and the sky and the south wind.

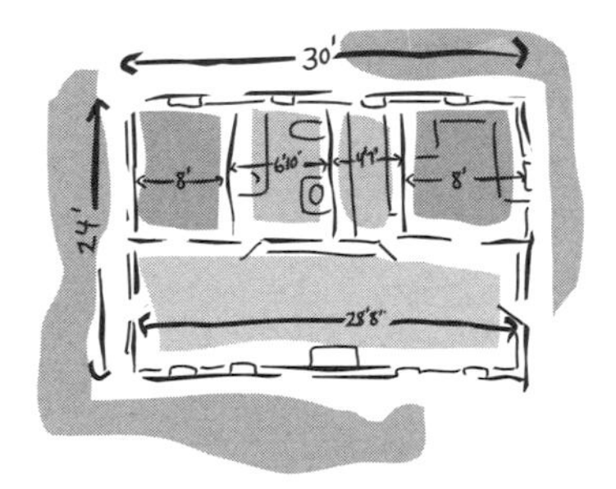

IMAGINING THE HOUSE

Imagining the house I would build for myself occupied idle moments of my senior year. Starting always with a rectangle and then filling it in, I must have exhausted the possibilities with my sketches. And at night I would live in that house to see how the floor plan worked out, the shelf space and storage. It would have to be a small house, with a place for everything, like a boat I was going to sail around the world.

Most weekends of my senior year in college I spent out on the land—replacing posts in the fence, adding another wire, hanging a gate, and clearing the building site. Days I was driven inside by the weather I spent working up construction drawings and materials lists, paring a dollar wherever I could.

I had never built anything on the order of a house, but I liked to read and put great faith in books. They taught me all I needed to know about masonry construction and roof framing, enough to get started. The floor plans though were all for houses on city lots along suburban streets or else for vacation cottages at the lake. The rooms of these small houses were dark and confining, and there was too little storage and work space.

In the end I went my own way. The twenty-four by thirty foot house I designed for myself provides for a large dining room/sitting room/library, half the size of the entire structure, fronting south with four large windows for summer and a woodstove for winter. The north half, the dividing partition also supporting the roof peak, consists of a corner kitchen with lots of work space, a large walk-in pantry, a bath, several closets, and a small corner bedroom. With its high ceilings and openness and white stuccoed walls, I like to think it is manorial and at the same time a peasant farmhouse.

First, before I could build, a well had to be proven. There was an old one on the property so full of roots I couldn't plumb its depth. But at least there had been water. A mile east is a fault line, limestone, and there isn't a drop. A mile north and the water tastes of coal. And even here, where I stood, the water sands are reputed to spread out like the fingers of a hand. It was very much hit or miss.

The three days it took to drill the well were days of suspense. Drilling a well is like nothing so much as digging for buried treasure. The rig, an old cable tool, older than the man who operated it, should never have worked at all. Vibration is the enemy of everything mechanical, and yet here was a machine that worked on the principle of a hammer and chisel, raising and dropping a heavy drill bit at the end of a cable until a hole is driven through the rock. Above the racket, the well driller and I conversed in shouts, the shuddering rig threatening to tear itself to pieces and sometimes succeeding.

At forty feet I could see the first water trickling into the hole, more at eighty. At a hundred feet the driller and I decided to stop. Somewhere there was a seam of coal, just waiting. And lead and zinc, which I didn't like the sound of.

When the bailer was pulled up out of the hole for the last time, we drank a toast from a coffee can. The water was clear and cold, 59° F, a temperature it would maintain the year round.

That done, I next had gravel laid for a drive, following traces of the original clinkers as far as they went, in a flat S-curve ending in a space wide enough to turn in without backing, about a hundred feet in all. Once a pole was set for the electricity I was ready to begin—first building a six-by-eight-foot structure above the cellar steps, so I could enter by a vertical door, and then a five by six foot pump house, at the same time teaching myself to lay cement block. Stuccoed inside and out, this form of construction is immensely solid. In itself that did a great deal to build my confidence.

I built my house in the heat of summer, working from first light until noon. My father lent a hand with the two-man jobs, and I had some professional help with the foundation footings, wiring and plumbing—but the rest I managed myself, learning as I went along. In the cool of the evening I cleaned up the site and laid out materials for the next day. Nights I slept in a small camping trailer. Work went well, the days measured in the number of blocks laid, 1,058 before I was done.

The hardest job and the slowest was the reinforced bond beam running right around the top of the four walls, carrying those buckets of wet concrete up the scaffolding. Most enjoyable was troweling the stucco onto the completed walls. Until then the house had looked depressingly like a laundromat. Now it was so classically white and pristine in the moonlight that I moved in, unrolling my sleeping bag in the roofless house. Those nights were unforgettably perfect. I can remember thinking it was almost dangerous to be so happy.

July was the hottest on record with an average high temperature of 103° F, but because of the drought the nights were cool. So that the stucco wouldn't dry too quickly and crack I had to keep the walls damp, spraying them periodically with a hose. The merest whiff of wet plaster brings it all back with a rush.

—Wet plaster and the smell of pine, raw pine lumber. With the four walls of the unfinished house blocking the lights around, especially the hideous security lights, dark night was for a time restored. I could hardly sleep for watching the westward procession of constellations overhead, Corona and Lyra and Sagittarius, the bright star Vega threading her way through the bare rafters of my house.

Construction is one of the best times in the life of a house, and I was glad not to have missed it. From the first our lives were inextricably mixed, mine with that of the house. At night from where I lay on my bedroll I could see by how many courses the walls had risen that day. And by the time I was sleeping on a bed in a room I had the feeling I had always lived there.

When I get up at night to see why Berenice is barking, it never occurs to me to turn on a light. Even with my eyes closed I know where everything is. I put it there—first as an idea, then as a pencil drawing, finally as a wall or a window—laying the blocks and framing the opening. The same could be said for all of Southwind Farm, the outbuildings and gates and paths.

The dark I always think of as friendly, because I know where I am.

The house never had a chance to be new or unfamiliar. It was after all built by an amateur with fairly rough materials. The four main interior walls are natural stucco, the partition walls of varnished plywood with mouldings of cedar one-by-fours. The ceiling is also of plywood, painted white with cedar strips at the joints giving an impression of beams.

The room in which I do most of my living is airy and lofty, and at the same time congenial. At one end is a small rolltop desk and bentwood chair, a large fruit press, several tall potted plants, and an old couch with an Indian rug covering a worn spot in the upholstery. Along the south wall are four windows and a woodstove, overhead in the center of the room a ceiling fan, and on the north my bookshelves and lamp and reading chair. And at the east end of the room is a chest of drawers and a dining table and chairs—where I am sitting as I write.

The arm of the chair on which my elbow rests is typical of everything in the house in having a history. The chestnut brown enamel has worn through to the darker brown below, both my brushwork. The ring of green below that and the underlying maple varnish are the previous owner's. And the grain of raw wood, glossy from my elbow and most beautiful of all, is the tree's. Someday I intend to strip it all down to the bare maple; meanwhile there's the fascination of that chair arm.

This is where I sit at mealtimes. Outside, in the poultry yard beyond the window, one of my hens is taking a dust bath, fluffing herself up to an astonishing size. A jumping spider, my indoor species Phidippus audax, is stalking a fly on the window pane. Even the stucco walls are a source of passing interest. Here the sweep of my trowel has left behind what looks to be a question mark and there, above the radio on my dining table, a map of the North American continent, its mountain ranges standing out in bold relief.

Cleaning house is a particular pleasure. I wouldn't trust anyone else to do it, not that that's likely. The nest of a chimney swift pinned to the bookcase is so fragile that the slightest brush would reduce it to sticks. And only I know the cracks where each of the jumping spiders lives, whom I know as individuals with distinct habits and personalities. The scent of freshly mopped floors, of wood smoke and creosote (pyroligneous acid) from cleaning out the stove, of pine woods from the exposed two-by-four studs in the open closets is indescribably pleasant.

The house has taught me to appreciate the fragrance of things, though describing them is another matter. The scent of a leaky honeycomb in the pantry is "mead-like," that of the grain mill is "a lingering sweetness," and the fruit press in the corner of the living room is "high summer." That's the nearest I can come.

The same with the sounds of the house, which are also the sounds of the seasons. The floor joists, wall studs, and rafters creak with the rapid temperature changes of spring. In summer I'm often tricked into believing a light shower has been more than that by the roof runoff splashing into a rocky puddle beneath the Russian olive. In fall when the first good rains come on a strong northwest wind, the gurgling and splashing of the gutter outside my bedroom window sounds like laughter. In winter my woodstove pings and squeaks contentedly when it's got a hot fire inside—and complains when the embers are beginning to cool, snapping sharply at me to demand more wood, waiting until I'm nearly back to sleep and then snapping again. The electric fan is not nearly as theatrical but is just as expressive, its somnolent drone the most characteristic sound of a summer afternoon—suggesting the buzzing of a mud dauber wasp at work or the whisper of a light breeze through the screen of an open window.

Appropriately, a mud dauber has built her nest in the air-conditioner, which I won't use until canning season, if then. In a small house with large windows and a high ceiling, bare floors and masonry walls, the heat of summer is better than shutting the season out. Having scheduled my work for the early morning and the evening, I sometimes read on summer afternoons or just lie still and listen, the call of a mourning dove deepening the silence.

Even in winter the life of the field finds its way in. One January I was wakened by the frightened chirpings of a least shrew, the repeated two-second trills strangely penetrating and musical. The sowbugs had been on the increase, wintering under boards or wherever there was any cover, and on a foray the tiny shrew found itself stranded indoors. After admiring its sleek coat, grayish-brown with a golden cast, I turned it out where I knew the hunting was good.

Others have come too, not as welcome as the least shrew. Once my house was broken into, my losses nothing of particular value—small change, a roll of stamps, my childhood penny collection I hadn't kept up, a pair of binoculars, inexpensive and replaceable.

The cats came out of hiding to tell me what happened while I was away. And together we put things right—my work clothes back into the closet, my journal on the desk, my books in the shelves. What did they make of the cecropia moth cocoons or the bamboo cricket cage? Or the straw skep or my honey suit, white coveralls stiff with wax and propolis? Or the entry in my journal for August 28, which is where the notebook had fallen open, about the dragonflies I had seen that day, various skimmers and darners—all so meaningless to anyone else.

And the same is true of the house. Small, eccentrically planned, hopelessly homemade, without central air-conditioning or modern heating, only one bath and no dishwasher, washing machine, or dryer—a real estate agent wouldn't know what to do with it. What price could he ask for the way footsteps echo? Or the way sunlight falls across the floor just so?

Yet if anything should happen to the house, I could rebuild it all from memory. Not in one long summer, though. Too much has been added since, along the east front, a large screened-in porch with table and chairs, a box for my boots, Berenice's house, and a porch swing. —And if anything should happen to me, whoever comes to live here will find himself falling into the habits of the house and taking up my journal just where I left off.

JOURNAL: MIDWINTER

(late December to the end of February)

December 30*

Change with little net difference, stratocumulus mostly cloudy later with cirrostratus then stratus overcast in evening, 64° to 73° suddenly falling to mid 40s in evening, moist air, moderate SSW wind decreasing and switching to gusty N wind. / Worked some on garden plan, gathered kindling down at the old corral (mostly black locust), inventoried honey at year's end. / Trunks of black locusts splotched with a distinctive, woody, hoof-shaped, and apparently long-lived polypore bracket fungus, probably a Fomitopsis, "locust forms" (once used as a medicine for horses), resembles F. pinicola, "redbelt"; pore tubes rusty and minute; cap woody and ridged with growth rings, some green with algae and colonized by lichens; interior layered with palisades of pore tubes, each probably representing a year's growth, lustrous golden brown with a hint of copper—and beautiful as a semi-precious stone.

*[*Journal entries are a sampling of days drawn from the thirteen years I have lived at Southwind.]*

January 1

Change to Fair and moderately rising, cirrostratus cloudy becoming clear, 11° to 17° F, moderate humidity / Cleaned cow stall, got my weekly winter bee sting, made butter. / Tonight, air is breath-catching cold, atmosphere seems to have vanished into space, long moon-shadows.

January 10

Change and rising very slowly, cirrus and altostratus mostly cloudy then clear, 20° to 55° F, dry air, N breeze. / Substituted at middle school (eighth grade science), called church trustees about scheduling a meeting, long reading evening. / Pond "warbling" tonight, short twitterings like the tinkling of spring peepers—from cracks forming in the thin ice of the pond, cooling after a mild day.

January 12

Change and rising very slowly, clear, 31° to 68° F, moderate humidity, light S to W breezes. / Sunday school and church, letter to Bruce [my brother], sold out of milk. / Late this afternoon the landscape shimmered with bright threads of sunlight, the silk of dispersing spider-lings. Even the waves of wind seemed visible.

January 16

Rain to Change and rising rapidly, dense fog becoming nimbostratus, mostly to partly cloudy before closing in* *with very dense fog after nightfall, 33° to 41° F, damp, NE to ENE breeze. / Began watching for cow's next heat, repaired bee equipment (indoors), ground flour and made bread. / This afternoon found several small waxy golden brown toadstools on open ground, ¾″ diameter on 1″ stems, quick growing enough to fruit even in January.*

*[*But not necessarily rain—Rain, Change, and Fair being divisions on the barometric scale used to broadly denote air pressure.]*

January 18

Rain and falling moderately, cirro-cumulonimbus and altostratus, fair to partly cloudy, then stratocumulus becoming nimbostratus cloudy, 33° to 57° F, moist air, NE breeze becoming light to moderate E wind then S breeze, evening thundershower (.27″). / Worked on bee equipment, made zucchini soup, mixed up some rolls. / Moist earth smell today, neither leaf decay of fall nor green of spring, simply earth. Winter flora like little green stars: shepherd's purse and pepperweed (mustards), beggar's lice (wild carrot), sowthistle, and henbit. Winter annuals a strong green, especially rescue grass and winter wheat.

January 24

Rain to Change and rising moderately, clear becoming cumulus fair developing into snowy cumulonimbus and stratocumulus mostly cloudy then clear, 23° to 49° F, moderate humidity, WNW breeze briefly gusty as squall line passed then light WNW breeze. As cumulonimbus squall line passed over late in afternoon, it was accompanied by intricate curtains of snow-virga, dark gray in shadow and burnt gold in sunlight. / Potted up some seedling trees, cleaned plow, fixed gate, planted apple. / Visited the white-footed mouse. Discovered her nest ball of fine grass blades and rabbit fur when moving one of the beehives. The native mouse is a good housekeeper, her habitation very clean and quietly dignified.

February 9

Change and nearly steady, clear becoming filmy, cirrus fair, 28° to 52° F, moderate humidity, NE breeze becoming SE. / Saw two streamlined and bushy-tailed coyotes moving across haymeadow to the north on some silent errand, probably returning from a night's hunting to the west. Saw another in bottomland clearing this afternoon, long and coarse-textured fur made up of the tans of grasses and the grays of twigs. Catching sight of me, it purposefully passed from sunlight into shadowed weeds.

February 12

Very Dry to Fair and falling rapidly, cirrus and filmy

cirrostratus mostly to partly cloudy, 7° to 37° F, dry air, light SSW wind. / Got feed in town, made bank deposit, cut mulberry roots growing under raised garden bed (heavy digging, chopped roots with hatchet), made soup for dinner. / Today the snow blanket thinned enough on cow pasture for regularly spaced tufts of bermuda to show. Late afternoon sunlight stained the tufts to copper straw, a striking contrast with the cold blue snow. And today filled with the clear warbling of meadowlarks. Contrary to what some say, they sing year round and not just in the spring. I have even heard them at night. The yellow breast feathers of the meadowlark are really of three distinct colors, yellow tips with a white band across the middle and gray at the quill end.

February 15

Rain to Change and rising rapidly, clear, 35° to 55° F, moderate humidity, light to moderate NW wind then light S air in early night. / Substituted half a day, made bread, removed mulch from the herbs. / Small puddles in cow paths swarming with animalcules visible to the unaided eye: 1) Cyclopoid Copepods with transparent body made rusty pink by many red vessels, two very large egg bags at base of body near tail, each with maybe twenty dark gray eggs; 2) transparent Monostyla-like Rotifers, smaller and more numerous than the Copepods.

February 25

Change and falling very slowly, nimbostratus fog then clear, 29° to 54° F, moist air, S breeze. / Spaded up raised bed, leveled recently plowed garden, mulched drainage ditches. / Up close, spring peepers produce a liquid metallic tinkle, a very clear and penetrating sound that loses little of its intensity over distances less than about one-half mile. They are extremely sensitive to movement (ground vibration) and must also see quite clearly at night. What sounded like hundreds of peepers at Little Marsh turned out to be two or three dozen, mostly clinging to bushes or trees standing in their shallow water habitat.

The garden year begins . . .

The garden year begins in the fall, especially when the soil has lain fallow, last year's harvest having been field weeds. As soon as the house was habitable I took time away from construction to lay out a garden. Even before deciding where to put the house I had picked a site, the Old Dutchman's barnyard. Here the soil was good, fertile and friable, dark and easily crumbled between the fingers. And the subsoil was deep, clay shading into mealy sandstone with shale below that, a good water reservoir.

The soil here is classified as of the "prairie" type, the other major grassland soil being "chernozem" (Russian for "black earth"). While not as deep or rich as that of Iowa, it is nevertheless among the most fertile of global soils. Forty miles northwest is the Tallgrass Prairie Preserve, the Osage Hills of Oklahoma and Flint Hills of Kansas, together comprising the largest remaining tract of tallgrass prairie.—

And had this twenty acres of mine never been plowed or pastured the big bluestem would still stand as high as eight feet, in places ten.

As it is, even the weeds growing here are a favorable sign—lambsquarters and giant ragweed and pigweed, a wild amaranth. Most of this prairie soil is slightly acid, nothing that can't be fixed with an application of lime. Wiregrass, on the other hand, is a sure sign of highly acid, clayey soil. Curled dock and wild carrot suggest poor drainage—and so do "crawdad chimneys" left by crustaceans which live in the meadow, sometimes hundreds of yards from any open water.

Common as crawdads are, they always look out of place, too much like miniature lobsters. And too beautiful to account for in a creature that lives mainly underground, dark greenish-blue trimmed in orange and cream and enameled to a high gloss. On spring evenings after a rain I have to watch my step in the pasture to keep from treading on them. They warn me off by waving their claws in the air at my approach in an effort to appear menacing. It is menacing if you're about to put a garden there, where the subsoil is mud!

Here on the site I picked, a gentle east slope would provide for runoff and reduce the sun's heat during the late afternoon. And the lay of the land would offer some protection from the prevailing summer winds, at least until I could get trees established.

So I staked out a large garden, about a quarter of an acre in all, loosely squaring it in Pythagorean fashion, and asked a neighbor to break the sod with his tractor and deep running plow. Thereafter I would use a cultivator and a hoe, mainly the hoe.

That first fall I planted winter wheat and the following spring forage soybeans, as a cover crop to shade out traces of bermuda grass and as green manure when it had been mowed and turned under. If it hadn't been for the bermuda I might have been tempted to plant a garden that first spring. Instead, I settled for a few tomatoes and other easily managed vegetables and spent the season working up the soil in the main garden—spreading and cultivating in composted manure, wood ashes, bonemeal, and ground limestone. At the time I was disappointed about getting off to what seemed a slow start. Now I know it was for the best, and all because of those ominous runners of bermuda. Something like it has happened so many times, a nuisance turning out well in the end, that I've learned to be philosophical. In gardening it always pays to be deliberate.

With the hand-guided rotary plow I cut the garden area up into six uniform sections, long rectangles thirteen feet by sixty. Each is plowed separately, raising the bed itself and leaving a drainage ditch between. Over the years I've added so much organic matter—composted manure and straw, green cover crops, the previous year's growth—that the garden is six inches higher and the soil light and crumbly.

Most of this work was done on those bright, cool days of winter when I was glad for something to do out of doors. And these fertile, loamy raised beds have every advantage: warming and drying out earlier in the spring, retaining moisture better in a drought, aerating the soil, making it easier to sow the seed and weed the growing plants, and producing far, far more. My garden is in fact absurdly abundant—but then I raise a great many things for the fun of it, because I like diversity and plants with difficult reputations, like English peas and figs.

Some years the weather of the Southern Plains can be difficult. But in recent years it has been nearly perfect, with springs cool enough for English peas, spinach, head lettuce, radishes, and beets ... early summers mild enough for potatoes and cantaloupe ... and late summers hot enough for okra and sweetcorn, and yet not so hot the tomatoes won't set ... with rain at almost weekly intervals. But I'm always looking into dry-weather varieties, against the day.

To some degree the climate of a garden can be altered. Mulching retains moisture … raised beds promote deeper rooting, equipping the plants for dryer summer weather … a soaker hose is the most efficient way of watering … and the taller crops can be used to shade and protect more sensitive ones.

The garden year has settled into such a pleasurable routine that I would like nothing better than to have it go on forever: plowing in the fall or winter, rotating the beds so that the previous year's dead furrow becomes next year's headland, planning the garden on the most dismal of winter days when weather has driven me in—and building up the soil on the dry, sunny days when I can work out of doors—cultivating the garden for a final time in the spring, staking out the crop locations, sowing, thinning, weeding, mulching, training the growing plants, watering with a soaker hose, harvesting …

The growing season is so long on the Southern Plains there is seldom a time when I'm not gardening. Spinach, lettuce, mustard greens, onions, beets, potatoes, Swiss chard, collards, radishes, and peas can be planted in the waning days of winter, though I've learned to be patient and wait for warmer soil and settled weather. Collards and Swiss chard, both planted in the spring, produce well into late fall, collards sometimes surviving the winter. Tomatoes sown directly in the garden in May come into peak production during late summer and fall. (Arkansas Traveler, a pre-1900 heirloom variety, takes drought and heat well.) Successive plantings of sweetcorn, a dozen or more through the season, will supply fresh ears right up until frost.

Fall radishes, planted weekly from late August to late September, are mild and crisp. Turnips aren't planted until about the first week of September. And even green beans planted as late as mid-August will usually ripen before frost. By late summer the compost added at the beginning of the season has been thoroughly broken down and assimilated. And the days then are so bright and clear.

Even in winter the harvest continues—spinach and kale from the greenhouse.

I can easily feed myself entirely from the garden (and did before I had poultry or a cow) well and at little cost. Regardless what kind of season it has been, here there is always an abundance. Though it produces no dollar income, the garden is my chief occupation and the basis of my economy. Here I reap what I sow, and the fruits of my table are the fruits of my labor.

And not just labor. To echo what the Koran says about life, gardening is "a sport and a pastime," a game played out of doors in the best of seasons with good competition—weeds, wilts and rusts, insect pests, dry winds and the capricious plants themselves. And the playing field is wide, wider than my garden—the planet and its rain clouds, the sun and (according to those who plant by the almanac) the moon.

Even the real work of a garden is hard to distinguish from its pleasures. Hoeing, which some consider the worst job, is no exception. And the swan-necked hoe is, in its utility and simplicity, one of the most beautiful tools. Mine has a small blade, forged and of one piece with the shank and ferrule, and a long ash handle. With it I can scrape the weeds off at ground level with an easy, sweeping motion. The secret is not in the hoe—it's in keeping up with the garden.

The Johnson grass is gone now, and the Bermuda is rare. Pigweed, lambsquarters, giant ragweed, and carpetweed are easily kept in check. Only the crabgrass, brought in with some wheat straw, is serious competition. And I'm content to hold a line against it, or to retreat only grudgingly. Since crabgrass is an annual, I can gain back each season what ground I lose by late summer. As long as the weeds aren't contending with the vegetable plants they do no harm and under certain conditions even do some good. There is something unnatural, even depressing in rows of weedless crops, just as there is in uniformly perfect fruit (as Snow White learned from the poisoned apple).

Most of the damage blamed on pests is really the result of bad gardening. Healthy plants can shrug off a great deal. I occasionally step in with one of the "organic" remedies—Bacillus thuringiensis (a bacterium harmless to higher animals) for leaf-eating insect larvae, diatomaceous earth (fossil diatoms) for soft-bodied insects, pyrethrum (derived from a kind of daisy), and rotenone (a tropical plant extract). Each does its job in ways the very strictest environmentalists approve. Tomato hornworms I simply pick off, the defoliated end of a stem giving them away. Squashbugs share my fondness for zucchini, so I make a point of hand-picking them too, along with any leaf on which their clusters of reddish-brown eggs appear. Since few adults overwinter, this interrupts their cycle, meaning there will be fewer next year.—But mostly I just plant a little extra for the insects and moles (and for the damage Berenice does digging for moles).

But the key to the garden is the hoe. Working my way quietly down the row gives me a chance to leisurely inspect each plant. Nothing can improve a garden's appearance so quickly. Weeding in good time and in loose soil is unaccountably pleasurable—steady and rhythmic, relaxing and congenially physical, leaving the mind free for reverie.

The garden becomes a landscape with "woods" of sweet corn and "savannas" of newly planted crops, okra and tomatoes, and "open plains" of recently harvested ground. It is "valleys" and "hills" of squash and melon, and in dry weather when a hose is trickling, "springs" and "watercourses." A silly sort of daydreaming, I suppose, but it taught me how to observe, reminding me that there are other worlds.

By June half a dozen species of jewel-like flies, from tiny iridescents to large bluebottles, have taken up residence along with at least as many kinds of ants, some barely visible against the soil and others (like the carpenter ant) large and robust, each intent on some pressing business. And there are numbers of hairstreaks, butterflies so small they're overlooked outside the garden, beautiful with finely patterned wings—and other butterflies, bold coppers and tiny blues and yellow clouds of hovering sulphurs. Ladybugs, both adult and larvae. And bumblebees on pea blossoms.

A lizard, a six-lined racerunner, took up with me, and through several seasons was company while I hoed. As was the killdeer who made her nest there, laying four speckled eggs. She so enjoyed carrying on, pretending to have a broken wing and shrieking hysterically, that even after we had got quite used to each other she insisted on going through her routine like any artist, adding still more extravagant embellishments.

A tiller would have overwhelmed all that, blotting it out. I do use a small one to cultivate in composted manure and a large wheeled mower to reduce corn stalks and tomato vines to mulch. And I have access to a hand-guided rotary plow that I bring in once a year, in the fall or winter. But most of the time I keep the machine out of the garden. As much as I respect what it does on occasion to make my work easier, the machine draws a veil between the mind and the hand. And I prefer to get by as much as possible with the simplest of tools, a hoe.

For much the same reason, I would rather plant nonhybrid vegetables—because it allows me to save seed from the best plants, thereby developing varieties better suited to this climate. Hybrids do not breed true, the offspring reverting back to various combinations of the two parent lines crossed to produce the hybrid generation. Despite the claims, I'm not convinced hybrids offer any advantage over a well chosen nonhybrid. In the garden, for example, many of the red-fruited tomatoes turn out to be surprisingly similar, though I am always on the lookout for any variety (like Arkansas Traveler, a nonhybrid) that is outstanding in flavor, longevity, and disease resistance.

To maintain vigor and genetic diversity I make a point of saving at least five hundred seeds from ten per cent of the plants. Through selection I appear to have achieved some improvement in several varieties. For example, I've selected to improve my sesame in two ways. By saving seed from the bushiest plants, I've reduced late-season "lodging" (collapsing under pressure from wind and rain). I also select for plants which bear the sesame seed in more densely packed ranks around the stems. In Alamo-Navajo Blue, a variety of flour corn developed by the Indians of the desert Southwest, I've achieved a noticeable improvement in ear blockiness and plant productivity through selection. And in okra, working with a variety called Clemson Spineless, I've improved both plant longevity and bushiness.

In the case of most crops, however, plants are so uniform it's difficult to select for particular features. Here I concentrate instead on selecting the best variety, looking through the seed catalogues for those which sound promising and giving them a try. Over the years several have become permanent residents for which I have a particular affection—like the Maestro pea, South American Giant popcorn, Butternut squash,

Dark Green zucchini, Marketmore cucumber, a grain amaranth with the arcane sounding name of K432, and the Kennebec potato, from which I've been saving seed potatoes for more than ten years.

From all the legumes it's also quite easy to save seed for the next season—mung beans, shellbeans, bush green beans, and the pole beans I grow on tall teepees made from saplings. Persimmons, which need to be thinned anyhow, are best because of their rough bark. Grown in this way, pole beans take up less space and are attractive standing in the garden, productive and easy to pick.

When saving seed there is a certain amount of selection that goes on quite unconsciously, and with it some improvement—a meatier butternut squash, for example, because I've naturally been saving seed from those with the smallest cavity. I'm trying again with a handsome variety of watermelon, an improved Moon and Stars. It's a dark green melon, speckled and spotted. And every season I try something new, this year a perennial onion and a black-eyed pea by the name of Running Conch and an upland paddy rice, which ironically it has been too wet to plant!

It's surprising how much personality a variety or even a particular plant will take on once you get to know it. I don't consider myself any more sentimental than the next, but in the course of those quiet hours in the garden with my hoe an affinity develops, one that deepens immeasurably in the case of seed I save back for planting next spring. Because of the seed in those jars and packets in the pantry, winter never seems long. Brought out on a snowy evening and scattered on the dining table in a pool of lamplight, seeds look like nothing so much as precious stones—treasure, which in a real sense they are.

JOURNAL: SPRING
(March and April)

March 8

Change and rising very slowly, nimbostratus mostly cloudy becoming fair then cirrus fair, 53° to 74° F, moist air, breeze SW then WNW then variable. / Finished working on taxes and county business assessment, picked up some gallon jugs for selling milk from Wilson School Cafeteria (gave the cooks some honey for their trouble), planted peas. / This evening almost roaring with the song of crawfish frogs from every pond. Spring peepers, leopard and chorus frogs all vocal, too, the chorus frogs with their rapid clicking. But most musical of all are the American toads, each "whirring" on a slightly different pitch. The night sweet with moist-earth fragrances and bright with star-speckled moonlight.

March 16

Change and falling moderately, cirrus and altocumulus fair then cirrocumulus fair in evening lowering to altocumulus, 40° to 70° F, dry air, strong SSW wind. / Sprayed dormant oil on fruit trees, worked with bees (medicating, feeding several hives, uniting two weak hives and equalizing brood in several others), began spreading rotted manure on garden. / Last night's northeast breeze heavy with the scent of woodland smoke, a rich and earthy pungence; today's southwesterly winds brought the sun-struck haze of grassfires, a delicate fragrance distinctly sweet.

March 23

Change and falling very slowly, clear then altocumulus fair early night, 30° to 59° F, moderate humidity, breeze N then S to SE. / Watered early garden, reviewed tractor maintenance notes, worked with bees (medicating, checking queen quality, removing entrance reducers). / Chorus frogs sang all day from pools on the poorly drained upland. Song peaks at dusk and is concentrated at one pond for a particular evening while others remain nearly silent, shifting to another the following evening. Yesterday it was Northwest Pond with crawfish frogs ascendant. Tonight Little Marsh is roaring with both Northwest Pond and Southwind Pond contributing a little leopard frog laughter.

March 26

Rain to Change and rising moderately, clear then smoky haze late afternoon, 42° to 70° F, dry atmosphere, moderate N wind. / Painted tractor shed, pumped water out of cellar. / Distant woods are taking on the colors of lichens, muted golds and rusts and delicate pale greens. The scene is, in fact, strikingly similar to those miniature landscapes on lichen covered rocks.

March 28

Change to Rain and falling slowly, stratocumulus and diffuse cumulonimbus mostly cloudy then stratocumulus fair with cumulonimbus building across north and northwest—later slipping off northeast. 51° to 77° F, moderate humidity, moderate S wind, .02″ from morning thundershower. / Cultivated garden, staked locations for planting, ground cereal. / Thirteen days ago a scattered grouping of maybe 150 minute black eggs were laid on some clothesline laundry; indoors today they hatched into 1/20″ long worms (black head, body brown shading to tan at the tail). Membrane of hatching eggs transparent, worm emerging through lateral slit. Slender with spike-like hairs, two or three pairs of prolegs at posterior end, tending to arch like inchworms. Definitely Geometridae moth larvae. Hundreds of Wodehouse's toads croaking at sunset in the meadow north of the house—so many together they produced a continuous hollow buzzing.

March 29

Rain and falling very slowly, stratocumulus overcast becoming partly cloudy, then cirrostratus overcast moved in from the west, 55° to 75° F, humid, light to moderate S wind. / Stretched more barbed wire along a short length of fence the cow had loosened leaning over to graze, pulled some garlic that had become a weed in the garden, weeded asparagus. / Cottonwoods and poplars are casting their huge bud scales as catkins burst forth, the scales resembling beetle elytra or even seashells.

April 1

Change and rising slowly, cirrostratus mostly cloudy becoming cumulus fair then clear, 36° to 49° F, dry, light to moderate SW wind becoming moderate N wind by way of west. / Weeded and thinned in early garden, spread old sheets on some tender crops (freeze expected tonight), worked on Varroa mite detection boards for bees. / Tonight, beaded chains of red-orange mark distant grassfires in the Osages. This afternoon's windshift brought in a fragrant haze of smoke which turned the sun flaming orange.

April 8

Rain and rising slowly, stratus cloudy then clear early night, 46° to 50° F, moderate humidity, light WNW wind. / Fellowship dinner at church (took rolls and honey), went walking to see the progress of the season, read right through the afternoon. / Spectacle-pod (Mustard family Cruciferae) blooming with a pungence almost sickening. It, along with false garlic and bluets, are three intensely sweet-fragrant early wildflowers, also honey-sweet Antennaria (Composite family).

April 9

Change and steady, cirrus mostly cloudy becoming clear, 42° to 66° F, moderate humidity, N breeze. / First planting of sweetcorn in garden, began saving hatching eggs for incubation, moved house plants out to porch. / Found a jewel-like dung beetle (Phanaeus vindex), its iridescence striking: bright green wing covers (elytra), coppery-red thorax with lemon-green sides, copper head with satin black horn; all except horn rough-textured (which enhances iridescence), smooth black and green legs and underside—also very shiny and with yellow shading.

April 12

Change to Rain and falling slowly, stratocumulus overcast becoming stratocumulus-cumulus partly cloudy then mostly cloudy in evening with some thundershower cirrus above the hazy clear, 55° to 81°, humid, light E wind becoming SE switching to moderate NNE wind in evening. / Substituted (nearing the end of my annual limit of thirty-five days), mowed yard. / Several dozen paper wasps (Polistinae) emerged from their winter hibernation quarters under the porch, most of which escaped from around the door but continued to hover around; this is a chestnut and muted gold banded species with amber legs and wings. There has been a profound change since the last cold front: landscape is so green it's dark even under a gibbous moon, trees so heavily clothed with leaves the lay of the land itself seems altered. The slightest breeze is enough to set the branches in motion, and the breeze itself is green-scented.

April 21

Change and nearly steady, clear, 46° to 67° F, moderate humidity, NE breeze to nearly calm. / Planted tomatoes, watermelon, and squash. Set up bait hive for swarms. Repaired hay rack. / Dense star-fields of bluets suggestive of puddles after a rain shower reflecting the sky. A good spring, well advanced now. Caught myself staring at the sparkling sunlight in a bucket of water and later just standing in the garden taking in the scent of warming earth.

With all its outbuildings . . .

With all its outbuildings the farm is like a settlement, a colony with each of us having our own shelter. Dog, cow, weaned calf, poultry, bees, and myself. What began as a house standing solitary in a weedy field became in time a farm, but not overnight. A farm grows by accretion, piece by piece, not by plan so much as by necessity, evolving in ways a biologist would understand better than an architect or builder.

There is always some construction going on—of a screen porch on the house, a shed for the tractor, a fence around the peach orchard, a pasture gate, a chicken brooder, a flail. No work on the farm is more pleasurable than this. It combines the fascination of solving a puzzle with the delight of making something. Driven indoors by a rain, I've spent the happiest of afternoons with tablet and pencil and square, laying out yet another project.

For some things there are plans in books or even the finished article on a store shelf. But usually not—because they won't do the job, or because they cost too much, or because you've got materials lying about to find some use for, or because you can make something better, which you usually can.

The work is not exacting. You'll change it as you go along, or later. The result is a little makeshift, a little rough—although just because the wood has a nice grain you may find yourself putting a hand-rubbed finish on a milk stool that will spend its working life in a cow shed, flyspecked and spattered with manure. The important thing is it works well enough, it does the job. And even if it didn't there's no one looking over your shoulder, no next door neighbor to object. Building inspectors and zoning boards aren't interested as long as it isn't an eyesore or a hazard to public safety—which milk stools, like most of the things you make, aren't. You have only yourself to please.

And I have taken the greatest delight in those things I have built for my own use. Shortly after the house was finished—at least for the time being—I began work on a barn of the same stuccoed concrete block construction. Fourteen and a half feet by thirty-six, the building would house my bee equipment at one end (extracting tanks, hive building materials, and all the other paraphernalia of the honey trade), gardening tools in a center room, and, at the far end, on the other side of a concrete block partition, the cow. The proportions of this long, low building with its hip roof turned out to be so mysteriously pleasing that there must be an explanation somewhere in the canons of architecture.

The roof structure was so light and airy that I decided to leave the framing exposed. The rafters radiate from a point of the ridge board in a way that suggests the veins of a leaf or the spreading branches of a tree, an impression carried out by chance in the floor. There falling mulberry leaves left their imprint, a fossil from that autumn afternoon when the concrete slab was poured.

In the country the outside is always coming in. Large, burnt orange paper wasps winter under my porch and emerge on the first warm days of April to be trapped inside the screen until I let them out. The porch itself, nine and a half feet by twenty, I added some years after the house was built—enjoying the construction so much that I hurried through my gardening to get back to it and worked as long as there was light.

It's furnished with a table and two chairs and a porch swing, all of peeled white cedar. When the weather is warm I have my meals out there, the raised porch overlooking the farm. And on summer nights I spread my bedroll on the porch floor, where I can watch the prairie moon rise above the eastern horizon—and then the morning sun.

Everywhere I look I see things a professional carpenter or mason or architect would have done differently, and yet I had my reasons. It was with the occasional violent winds of the Southern Plains in mind that I decided to do away with any projecting eaves. The modified hip roof looked so French that when the natural stucco began to streak from weathering I decided to paint it in the manner of small houses in Provence—rosy terra-cotta walls with white corner detail, white borders around each of the windows, and Mediterranean blue shutters and door.

The crawl space under my house is deeper than usual making it easier for an amateur to work on plumbing and wiring. And entry is by a largish trapdoor in the floor of my pantry rather than the usual cramped opening in the foundation wall. The house may be structurally stronger for it—and replacing the elbow joints in my plumbing, a fault of the manufacturer, took only an afternoon. In construction at least, what is done for a practical reason usually turns out for the best.

In both the short run and the long, economy is also a consideration. The large windows, high ceilings, cool masonry walls, and screen porch enable me to live without air-conditioning. A fruit press was too expensive, so I made one. A small threshing machine, one that will winnow the chaff from the grain, is simply not available. And that is my next project.

There is almost nothing an amateur working alone cannot do, from building a house or a barn or a shed to stretching fence and hanging gates. And pitted against his constructive and orderly efforts are the familiar antagonists of a small farm—age, weathering, hard use by animals, and the consequences of altering a landscape.

A small farm is nooks and crannies, a toehold for rank nature. Trees take root, fencelines grow up in brush, the whole climate changes. Never before a problem in a dry land, rot takes hold of a shed's foundation and a trumpet vine lifts the roof off. Iron either rusts and flakes away in the hand or else takes on the patina of an ancient relic. Wood crumbles with rot or turns to brittle amber. Plants that before couldn't have been coaxed to grow on the prairie upland appear out of nowhere, take root, and go on a rampage.

In the shadow of even the neatest farm there is hint of disorder and creeping dilapidation. And the farmer himself—in his patience with inconvenience, readiness to make do, and reluctance to throw anything away—can easily fall under its spell. But that is not the same as "makeshift," which is one of the necessary arts. Making do with what is on hand.

The large basket I use for firewood, stacking the logs on end, has been reinforced so many times with sticks and wire that a visitor asked if it was "an antique withy basket." The chick brooder I made from an old metal garden cart. Discarded beehives (lids, bottom boards, hive boxes) are surprisingly useful around the place. If too rotted for anything else, hive boxes serve as rabbit guards for sapling fruit trees. An old mop handle I use to steer my chickens in at night. Two broomsticks I joined with a section of plastic pipe to make a long arm for my fruit picker.

Old boots are a handy source for a piece of leather or rubber, a hinge or a bumper. An old pump house cover I was about to haul off shelters a pair of five-foot constrictors (corn snakes, useful predators in the garden). A section of hollow sycamore by the barn hydrant (once housing a colony of wild bees discovered when a neighbor was clearing up after a storm) is a perfect work bench for cleaning gardening tools. There are even notches to brace the tools, a hollow to stash my "clettering stick," and a place to drape rags. I couldn't have designed anything better.

No dimension lumber ever gets thrown away. Every scrap of the original five-by-five foot covered porch went into something—the new screen porch, a shed for the weaned calf, shelving in the tractor shed, and the last of it into the woodstove as kindling.

In the extremes of makeshift, however, there is sometimes a surrealistic quality. Three or four miles from me is a pig farm. The tenant must at one time have had something to do with appliances, for he's made a windbreak of old refrigerators, washing machines, kitchen stoves. And he feeds his pigs on stale bread, truck loads of it, the loaves still in their wrappers, which the pigs and the wind scatter to catch in the fence—along with the feathers of peacocks. For strutting about with the pigs, in the spring spreading their shimmering fans, are a dozen or so peacocks. A mile beyond this pig and peacock farm is a housing addition, "Dover Pond," very grand and exclusive. I have not been down to see how the two are getting on.

In every small farm there is some degree of higgledy-piggledy. When I look around at Southwind—at the house and outbuildings and pens and fences and gates and beehives and all the rest—I'm surprised there isn't more disorder. I'm also astonished that it has been done so quickly and has given such pleasure in the doing. Having built the farm and its appurtenances, most of them, with my own hands has given me a heady sense of possession and permanence. It's hard to remember that all this was once an empty field, just as it's hard to believe that having been set in motion it won't go on being a farm forever.

JOURNAL: EARLY SUMMER

(May into July)

May 16

Change and steady, hazy cirrus and altocumulus partly cloudy becoming hazy clear, 56° to 80° F, moist air, light NE breeze becoming ESE breeze. / Sweetcorn planting #5, sprayed copper sulfate on grapes, checked the queens' egg laying patterns in the hives. Looked over mulberry trees along the west fence (may put up some berries, the peaches having got frozen). / The green meadow dotted with the strong primary colors of Indian paintbrush, coreopsis, spiderwort—and white fleabane.

May 18

Rain and nearly steady, cirrus and stratocumulus mostly cloudy becoming fair, then stratocumulus overcast moving in from southeast early afternoon, light S wind then moderate to strong SE wind decreasing and becoming southerly, .06″ misty rain in afternoon. / Mowed yard, made marmalade for Dad's birthday. / Evening sparkled silently with lightning bugs (Lampyridae).

May 25

Change to Rain and falling very slowly, nimbostratus cloudy becoming cumulus fair, then clear by evening, 64°to 86° F, humid, S breeze, .30″ rainfall from previous evening. / Serviced tractor, worked with bees (catching a swarm). / This evening air warm, flaccid and tropical, with insistent "clucking" of raincrows (yellow-billed cuckoos) and "squawling" of gray treefrogs—the raincrows feeding on hairy webworms in the mulberry.

May 27

Change and falling very slowly, cirrus fair, 57° to 83° F, dry air, moderate SSE wind. / Mowed brush (took lunch), early dinner, long evening for hoeing garden and reading. / Meadow wildflowers a contrast of cool blues (spiderworts) and warm yellows (coreopsis). Islands of poorly drained clayey soil are rosy-brown (lovegrass, the panicles in full bloom).

June 2

Rain and rising slowly, cumulus partly cloudy, 74° to 92° F, humid, SW becoming S breeze to calm. / Set potted trees in the old flat-bottomed boat for easier watering, took leaking tractor tire to be fixed, stopped by library, two quarts honey stolen from box out front (first time in months). / Rank growth on meadow knolls the aerating effect of moles? gophers? Their unused winter stores a source of soil nutrients? And fairy rings created by soil fungi more prominent this year.

June 7

Change to Rain and falling slowly, hazy clear, 59° to 90° F, dry air, moderate S wind. / Took stray dog in to vet's to be boarded until a home can be found, did some jobs for folks, picked blackberries. / At 2:00 a.m. a chorus of coyotes erupted in yapping just yards from my window. A startling sound even though I've heard their wild outbursts of song nearly every evening this spring and early summer.

June 17

Change and rising very slowly, altocumulus mostly cloudy giving way to cumulus and building up to thun-

dershowers (mostly cloudy), then cirrus fair by nightfall, 72° to 92°F, humid, light SE wind, trace of sprinkles around noontime. / Cleaned house, sweetcorn planting #9, set out some late tomatoes. / This morning watched a hawk crossing over my "woods" harried by nine scissortails and several blackbirds—the flycatchers swinging so wildly, tails streaming, they looked like gyrating kites in a brisk wind.

June 27

Change and falling very slowly, cirrus and stratocumulus partly cloudy early, then cumulus fair, 70° to 93° F, humid, light SE wind. / Dug potatoes (hot work), picked up 50 cases of quart jars at a discount for quantity. / An ideal summer day, in every way typical and perfect.

June 30

Rain and falling very slowly, altocumulus and cumulus partly cloudy, 74° to 103° F, moderate humidity, light to moderate S wind. / Mulched some crops in the garden, canned corn, very busy. / Found a black, ½" long-horned beetle—a convincing wasp mimic with jerky movements, quivering antennae, wide and flat femur (all three pairs legs), and white line segments on hard wings covering abdomen to give the impression of being "wasp waisted."

July 2

Change and falling very slowly, clear with slight dustiness, 71° to 102° F, moderate to low humidity, moderate SW wind. / Made butter, ground cereal (coarse) and flour (fine), made bread, a little garden work in the evening, Mom's birthday. / A diversity of summer insects singing these nights: the metallic buzz of angular-winged katydids, the rasping zip-zip of cone-headed grasshoppers, the trill of tree crickets, and the soft strum of ground crickets.

July 3

Change and steady, cirrostratus and stratocumulus cloudy becoming partly cloudy and fair by nightfall, 70° to 89° F, humid, calm then NW breeze switching to SE by evening, .18″ from morning thundershower. / Extracted first honey (about 600 pounds), bottled some for customers who called, must do some canning tomorrow! / A borer infested common sunflower (Helianthus annuus) is leaking frothy sap, attracting a great variety of insects: carpenter, harvester, and smaller ants; several large wood nymphs and question marks [butterflies]; blow, flesh and several other fly species; a honeybee; a flower beetle and many green June beetles; and on the underside of leaves are "herds" of aphids with their ant "shepherds." The sap has a bittersweet taste, weedy with a hint of woodiness.

July 15

Change and rising slowly, hazy clear becoming cumulus partly cloudy, then cumulonimbus mostly cloudy, then altocumulus cloudy, 73° to 96° F, moderate humidity, SSW breeze then light NE wind, trace of afternoon sprinkles. / Cut yelloweye shell beans, picked some winter squash, put dried herbs in jars, bottled some honey. / Gazelle beetles are back, this year fewer in number and feeding on potato leaves, 4 days later this year than 1984, 3 days earlier than 1983. Dates of their first appearance in recent years: 7/19/82, 7/18/83, 7/11/84, 7/15/85, 7/8/86, 8/26/87. Later in the morning, the herd of several hundred beetles of various sizes moved on to some adjoining sweetcorn—where the crackling sound of their eating was quite audible.

July 18

Change and rising very slowly, clear to cumulus fair, 74° to 96° F, moderate humidity, moderate S wind. / Mowed yard, extracted honey. / Herds of brown and dull yellow-striped blister beetles (Epicauta vittata) have been stripping wild amaranth weeds in the garden. Slender, active, and alert, they stampede for cover when one approaches. They live in colonies and undergo hypermetamorphosis. I call them "gazelle beetles."

July 20

Change and falling very slowly, altocumulus and cumulus fair with widely scattered thunderhead anvils late afternoon, 73° to 98° F, humid, calm then light ENE breeze in the evening. / Canned corn and tomatoes, flailed beans. / Summery: mown weeds and grass along north edge of garden dry and brittle, insects busy around the crowder pea blossoms (ants, flies, small wasps, and bees).

A dozen hens clucking
sleepily to themselves . . .

A dozen hens clucking sleepily to themselves through a summer afternoon is of all sounds the most characteristic of a small farm—that and at first light the crowing of a rooster, which can carry for miles. The call begins deep in his syrinx, the bird's vocal organ, also known as the "panpipe," where it seems to bounce back and forth before emerging as a rich, broad, horizon-piercing wail which finally thins into a distant echo. Passed along from farm to farm, west with the morning, I like to think a crow begun out on Cape Cod might carry to the Pacific. And on a clear morning even the silliest rooster can sound like a Roland or a Gabriel. But for all that I still prefer a hen's clucking.

The poultry yard is close by, near a south window where my dining table is. I sometimes linger over meals, reading or listening to short-wave. The chickens, though, are such a presence, so busy always at something that I find myself studying them with fascination.

Nothing escapes their notice, everything is worthy of comment. They are such expressive birds it's even clear what they are talking about—when the excited squawkings of a distraught hen refer to a passing hawk, and when to a quarrel over a grasshopper. And they are intensely emotional, harboring their fears and grievances long after the event—fretting about the hawk or lost grasshopper in a high, drawn out, scratchy complaint.

Laying an egg is such a personal triumph that the hen almost crows with conceit, the others chiding her in throaty undertones. I know from the fuss how many eggs I can expect to find that afternoon in the nest boxes.

The chickens have two fenced ranges over which to roam and a central scratching yard where I feed and water them and where they come running for their table scraps. Shaded by a hybrid mulberry, an Illinois Everbearing, the bare scratching yard is cratered from the hens' dust baths.

The intensity with which they fluff their feathers and immerse themselves in a cloud of dust is comic. But it is also serious business. Silica in the soil effectively pierces the soft bodied parasites and their eggs on the same principle as diatomaceous earth mixed with stored grain.

Given the chance a flock will keep itself in a fine state of health. I occasionally paint the roosts with nicotine sulfate to control body lice, and, during an unusual cold spell when the hens are under stress, add terramycin to the drinking water. For the flock's sake, the best medicine for an ailing bird is the hatchet, though this is rarely necessary if hens are not kept beyond a second year.

One of the pleasures of owning a flock is getting acquainted with the various breeds, all distinct and some of them quite beautiful. Earlier I had some slight experience with the Mediterranean breeds, Brown Leghorns and Blue Andalusians. They were excellent layers, almost too much so for their own health, but didn't dress out well. And they always seemed in a state of hysteria. The first breeds I tried at Southwind Farm were Rhode Island Reds and New Hampshires, both in the English class. Heavier hens which laid large brown eggs, they proved good all-purpose birds—but were almost as nervous as Leghorns.

The next, Buff Orpingtons, turned out to be good companions, and I am quite attached to the breed. Also of the English class, calm and heavy, layers of light brown eggs, Buff Orpingtons are noted for the fullness of their plumage. The hens are of a golden buff which fades somewhat as the summer progresses before molting in the fall; roosters are of the same buff coloring, though more lustrous.

He is a noble fellow, the Buff Orpington rooster, steady and alert, quick to warn the flock of danger. His full, deep crow carries authority. And he has the manners of a gentleman, calling his hens to a select morsel with a stuttered clucking and low hooting.

I picked up my first Buff Orpington chicks at a small farm deep in the country south of Oologah. When I arrived, the woman of the house, chickens scratching and clucking around her feet, was sitting in a straight-backed dining room chair in the shady yard, reading a book.

Sometime I might try a few Dorkings, an ancient Roman breed brought to Britain during the time of Julius Caesar. Among other features, they are distinctive for having five claws instead of four. Chaucer's Chanticleer was, I think, probably a Silver Gray Dorking.

Meanwhile I'm concentrating on my Orpingtons, introducing a fresh blood line every five or six years to reduce inbreeding, a problem with a small flock. Ordinarily two or three hens go broody in April or May, but I prefer to incubate. It's easy, and I can control the number of chicks, hatching twelve to fourteen every spring, more than can be reliably expected of one hen. There are also complications in letting a hen set—others wanting to lay in her nest. But if I separate the broody hen she'll be discontented and want to get back into the chicken yard.

Since I've not allowed my Orpingtons to become inbred, the incubator hatching rate is very good. After two weeks in the brooder, the chicks are transferred to a hut in a separate pen. Having grown accustomed to the sound of their peeping in the pantry, I find the house unnaturally silent.

I prefer to feed "chick grower" rather than "chick starter," which is always medicated. While I do treat their drinking water with terramycin the first few days, the routine feeding of antibiotics is a mistake. As soon as possible I get the young birds out on the range, where they can forage for themselves. I supplement the flock's range feeding with hen scratch, a few laying pellets during the height of the season, any surplus or sour milk, garden and kitchen scraps, grit and oyster shells. The fescue range provides greens all year for deep yellow egg yolks.

Chickens are happiest where they can roam freely. I have two large ranges, each of about twenty-five hundred square feet, between which I can rotate the flock. The life they enjoy is pretty much what it was for the original jungle fowl from which they descended, and their whole nature responds. My Buff Orpingtons have a lively intelligence and a subtle social order. But quite apart from the pleasure their company gives me, I consider sympathetic care an obligation.

From each hatching I keep two roosters and five or six hens, dressing the other young birds at about five months, in September or October. Of the older flock I keep maybe four hens for fresh eggs until the pullets start laying, giving most of the older birds to some family with young children. They are reasonably good layers for another year or two, and the calm disposition of the Buff Orpington makes it an ideal introduction to the keeping of poultry.

Slaughtering always seems to me inescapably destructive, the metallic golden plumage of the rooster losing its lustre before my eyes and the bluebottle flies appearing out of nowhere. For consolation I reflect on the fundamental honesty of doing it myself instead of leaving it to some industry. This isn't empty rationalizing on my part. The treatment of animals intended for slaughter, from the beginning of their lives to the end, is something for which we are accountable. That our only connection is with packages at the meat counter, done up neatly in plastic and barely recognizable, makes no difference.

Cockerels are aggressive and would eventually be

very hard on the hens. And surplus, aging, and ailing birds would introduce disease into a flock kept only for its eggs. Whatever I think of the job, there is no alternative to the killing cone and the hatchet. Once that is over the process of dressing the bird turns out to be rather enjoyable, and I always finish up with a sense of accomplishment.—There is in life on a small farm an element of coming to terms. But then that is the whole point.

Of far more importance to my domestic economy than the dozen or so chickens I put up each year, my sole source of meat, are the eggs. Hens given free range, where there is a ready source of calcium, lay tough-shelled eggs with rich golden yolks, the distinctive color produced by the same carotene that makes Jersey butter and autumn leaves yellow. And the flavor is fresh, with no trace of that pungency found in cage eggs—or for that matter in the meat of commercially raised birds. There is no argument for the industrial production of eggs and poultry that I couldn't silence by serving up an omelet or a drumstick from chickens given space to forage.

I can't explain the pleasure of gathering eggs from straw-filled nest boxes, but I'm not sure I need to. It's instinctive, something every child seems to understand. Perhaps the mystery lies in the egg itself, I don't know. A broody hen will fluff her feathers, hiss, and deliver a sharp nip. A hen simply laying with no intention of brooding a clutch will give me a somewhat startled look, and if I don't move cautiously and slip my hand carefully under her to get the eggs she'll bolt from the nest squawking bloody murder.

Freshly laid, hen eggs are covered with a waxy bloom, the creams and tans of Buff Orpington eggs, each one different, appearing to be enveloped in a satiny white haze. These are the eggs I gather each afternoon, always with a sense of expectancy. Egg laying in hens, just as with the queen bee in the hives, begins increasing shortly after the winter solstice and rises as the days lengthen.

JOURNAL: HIGH SUMMER

(late July into early September)

On the Southern Plains high summer is a "fifth" season—one of lofty skies, solid and intricate cumulus clouds, hot sunlight, bright stars (Scorpius and its companions), insistent south winds, and the singing of innumerable insects.

July 26

Change and steady, cumulus fair then cirrus and cirrocumulus fair, 67° to 87° F, moderate humidity, light NE wind. / Canned last of tomatoes, mowed some weeds in garden, flailed amaranth. / Several jewel-like leafhoppers (Cicadellidae) on Gold Coast okra: wings intricately mottled with fragments of sky blue (blue with a slight dull lavender cast) and blocky patches of golden orange (the color of oat straw) on thorax and wing borders.

August 2

Change and rising very slowly, altocumulus partly cloudy then fair with increased altocumulus and cirrus during afternoon and again by nightfall, 65° to 88° F, dry air, light NE wind, a few drops of rain early afternoon. / Made peach leather, watered trees out back (taking along a book), in the evening harvested blue corn. / Dryness lends air and the scents it carries a quality of cleanness; even the pungency of dust is clean.

August 14

Change and falling very slowly, moderate fog giving way to nimbostratus mostly cloudy, becoming hazy cumulus fair and hazy clear by evening, 59° to 95° F, humid, variable breezes. / Began checking hives for Varroa mite infestation, also assessed queen quality and colony populations. Ordered queens for fall introduction.

August 20

Rain and nearly steady, hazy clear to cumulus fair, 70° to 96° F, humid, light S wind. / Canned pears, ground flour and made bread, fed and watered neighbor's livestock. / These hot afternoons so filled with the metallic song of cicadas, continuous and unvarying, that one would think it was a sound made by the heat.

August 25

Change and steady, clear, 72° to 98° F, moderate humidity, moderate SW to S wind. / Began sun drying sliced okra, mowed brush, spent a long evening reading in the hammock. / Watched numbers of orb-weaving spiders building their webs in failing light of dusk. An Araneus spider began by stretching lines between tree and post then dropping anchor lines to the ground and stretching lines to branches above. Next the spokes were laid on the framework. And finally the sticky spiral web was paid out at a rapid pace. All was done very methodically, especially the initial framework which appeared to require some careful planning on the spider's part. The freshly-

spun orb webs glistened in the evening light, straining and billowing like sails with each passing breeze.

August 29

Rain and nearly steady, cirrus fair then cirrostratus mostly cloudy, 76° to 109° F, dry air, light to moderate SW wind. / Juiced watermelon for freezing, flailed more beans. / Air so hot it stings eyes, nose, and face—and is laden with a distinct scorched scent by evening.

September 7

Change and falling very slowly, clear with isolated cumulus morning and broken cumulonimbus bank in northwest evening, 70° to 92° F, moderate humidity, light to moderate S wind. / Fed colonies low on honey stores, sowed turnips and a second planting of radishes in fall garden. / Native pastures taking on the deep luminous pink of Gerardia (abundant this year), the azure of blue sage, and the rich burnished gold of Indiangrass (the plume-like panicles the nearest thing to sunlight). Night air heavy with the mushroom scent of August's dead grass when it is moist with dew.

September 13

Change and steady, clear to cumulus fair, 72° to 98° F, moderate humidity, light S wind. / First day of substi-

tuting this school year. / Late this afternoon came across a colony of pharaoh's ants in the process of moving their nest, a twenty-five foot column dense with workers carrying eggs and larvae moving one direction and returning empty-handed for more in the opposite. At one point along the way there was a loose cluster of maybe a hundred ants excitedly circling the queen in the center, but still moving toward the new site at a good pace. Could not imagine why these pharaoh's ants were abandoning the old colony—unless it was that the new location was more open and perhaps better drained, more suitable winter quarters.

There were already persimmons, mulberries, and sand plums . . .

There were already persimmons, mulberries, and sand plums when I came, growing in the fence rows, and blackberry bushes right to the edge of my garden. Until my own trees began to bear I gathered enough to have wild fruit every day, fresh or preserved.

Fruit trees are not finicky. They don't ask for much in the way of topsoil or climate, but they do insist on having good drainage, preferring a slope or crest of ground. When planting any tree dig a large enough hole that every root can spread to its full length, at the same time removing any grass roots from the loosened soil. After that don't cultivate the ground beneath or even mulch around fruit trees since this attracts gophers and moles.

I scatter wood ashes around the trees annually and from time to time a little agricultural bonemeal. Heavier fertilizing would produce faster growth but a weaker tree, vulnerable to insect damage and with branches easily broken by wind or under the weight of fruit. Pear trees which have been fertilized are far more susceptible to fireblight. Before winter I staple a sleeve of paper feedsack around the trunks of young trees to protect them from rabbits and sunscald. Otherwise the tending of fruit trees is mainly a matter of spraying and pruning, both in moderation.

Spraying is a bother, but not much. If it were true that you couldn't have fruit without using the more toxic sprays, then fruit wouldn't be worth having. In November I spray my stone fruits (peach, cherry, apricot, and plum) with a solution of copper sulfate and hydrated lime. This Bordeaux mixture, so named from the region in which it first came into use in the nineteenth century, is the original fungicide. Discovered accidentally when trees next to a road sprayed to settle the dust were found to be disease free, it is an old standby and works well enough.

In March, at the end of the dormancy period, I spray all the trees with a very light grade of miscible oil. Dormant oil spray, not a poison at all in the usual sense, works by suffocating insect pests and their eggs and has some fungicidal value as well. In April or May I spray the trees with micronized sulfur, a fungicide and insecticide in use since ancient times.

Also in April and May I spray for curculio, the brown snout beetle, my most serious fruit pest. In a year or two, when my can of Imidan is used up, the orchard will be producing so much fruit that the loss to insect damage won't matter. Imidan is a highly specific insecticide that's not particularly harmful to beneficial insects. And I spray only twice a year, just after petal fall and again two weeks later, instead of the ten or twelve times which is routine with commercial orchardists.

In summer I paint the fruit tree trunks with a solution of five percent Sevin dust to control borers, having lost more trees to that cause than any other. Moth balls buried at the base of the trunk didn't work.

If I had to use any more chemical controls than these I would do without the fruit, but never without the trees, which have always impressed me as having character.

I know an old apple tree, riddled with borers, that has been dying for as long as I can remember. In spring, limbs that won't survive the summer are covered with white blossoms. Some even set fruit, the small apples ripening weeks early because time is running out. And there is a pear standing all by itself in the middle of a wheat field south of here. No one ever picks the fruit, which doesn't matter to the tree. And the smooth, polished bark of apricot limbs has a more beautiful finish than any antique chair or Stradivarius.

When young, trees are pruned to develop a framework of strong fruit-bearing branches. After a good structure is established, I prune lightly to take out shoots and diseased wood and to keep the crown of the tree open to light and air. There are different schools of pruning. I prefer the modified leader system in apples, pears, and apricots ... the open-vase form in peaches and plums ... and the four-cane kniffin system in grapes. It really doesn't matter as long as you have some sense of purpose and method.

No job on the farm is more pleasant than pruning. I choose my time, waiting for a bright, still day in late winter, cool but above freezing. Etched against a cobalt sky, the upper branches are quite beautiful in the clear light—the twigs of the apricot a bright red, buffed to a high polish by the hard weather, and the peach twigs a light amber, almost the color of oat straw. I don't mind lingering over the work, making it last through several afternoons.

When pruning a limb leave the swollen "branch collar" at the base. This collar, the seat of the tree's resistance to disease, will close over the exposed cut so quickly wound dressing is unnecessary and sometimes a hindrance. And when cutting back a branch, leave as the new growing tip a bud that points in the right direction. All this takes a lot of sizing up from this angle and from that, but I am in no hurry.

Always I am struck by the individuality of trees, even of the same variety and parentage. The age of the tree has something to do with this, and its location—but I'm always left with distinctive habits of growth and fruiting I can't account for, except by attributing it to personality. Over the years I've become so well acquainted with certain trees, enjoying their fruit through the winter and watching for the first blooms along their branches, that I've ceased to think of them in strictly botanical terms.

The varieties of fruit trees are a source of fascination. Of apples, for example, the Red Delicious is well known but rather boring—until I discovered from watching wasps and bees that the slightly fermented juice is surprisingly nectarous. My Granny Smith, which didn't do well in Oklahoma, I grafted over to Maiden's Blush, an applesauce variety. And my Jonalicious I changed to a Kandil Sinap, grafting shoots of the new tree onto the rootstock of the old.

Apples are the most reliable orchard fruit, the trees long-lived and attractive. But the common nursery varieties are apt to be disappointing. So I began looking for apples that would thrive in the Oklahoma climate and learned to graft so I could propagate some of the more interesting and obscure varieties. The Kandil Sinap is from Turkey, and so is Anasya, the scion wood I'm grafting this spring originally coming from near Sardis, the ancient city of Lydia. Red Astrachan, which I'll graft onto rootstock I'm growing for the purpose, is from southern Russia. And the Yellow Horse graft, which is now a small tree, originated in the Old South. My Cummings, on the other hand, an unknown apple with great possibility, came from an early settler's abandoned orchard down the road.

The grafting I do is of the simplest and most traditional kind, either "whip and tongue" or "cleft," nothing anyone couldn't learn. All it takes is a grafting knife (razor sharp), budding strips (elastic banding to hold the cutting and rootstock together until the graft takes), and grafting tape (binding to add rigidity to the union). The scion wood, cuttings from the desired variety, is cut in late winter and stored in the refrigerator until April. More important than timing are ten days of settled weather after grafting the scion wood onto the rootstock.

It isn't difficult to learn from a book, but I had the advantage of knowing a self-taught botanist of remarkable ability, L.E. Scott, a specialist in wild fruits. As an elderly and dying man he was still taking cuttings he knew he would never live to graft onto rootstock, leaving instructions. His bed was moved into the living room where he could look out on his trees, cuttings of which are growing in orchards around the world.

But even without getting into grafting, the pursuit of just the right variety of fruit tree can be a consuming interest. The Yates bears small speckled apples with a scintillating crab apple flavor. Left on the tree until heavy frost, they will keep all winter. A favorite of mine, the Arkansas Black, is a heavy bearer of dark purplish apples with heavily waxed skins and flesh the color of Jersey cream. It's a hard apple, best when bletted by winter right on the tree. The Prairie Spy is an old apple, also late blooming. Yellow Horse and Carolina Red June are both new trees from the South. Mutsu, a Japanese variety, holds promise. And the Liberty is so disease resistant it may not require any chemical spraying, reason enough to plant the tree.

Whatever variety of apple you grow, provided it is grafted and not a chance seedling, is a living shoot from the original tree—whether the latest hybrid from Geneva, New York or the White Winter Pearmain, an English apple dating back to 1200 A.D.

Next season I intend to plant a Duchess of Oldenberg and a St. Edmund's Pippin, reputed to be the most beautiful of all the russets, a deep fawn color.

All grafted trees, not just apple, are in a real way quite old—especially some of the pears. Because of fireblight, I am limited to the resistant varieties, Delicious and Orient, the Kieffer (with its distinctive musky aroma) and the Seckel (also known as the Honey Pear). If the fruit kept better and wasn't so bland I would be even more enthusiastic, for the tree can be very hardy.

I had expected great things of my peach trees. The Bixby area, south of here in the Arkansas River Valley, is famous for its peaches. The Belle of Georgia, Summer Pearl, Burbank July Elberta, and Early White all tantalized me briefly with their fruit—the most delicious of all the summer fruits. But they proved too vulnerable to insects, disease, and late freezes to be worth the effort; and as I lose trees I'm not replacing them.

Even if every spring a freeze nipped the blossoms of my apricot, a Hungarian Rose, I would have a tree for its beauty alone.

Grapes are reliable, especially the American types—Catawba, Concord, and Sunbelt—even in heavier soil. Sour cherries will not tolerate heavy soil, no great loss since the wild blackberries are coming on at the same time. I used to raise strawberries but gave it up for the same reason—the blackberries.

I am fond of figs and cultivate a fairly large plot. For the lack of any other name I refer to the variety as Bayouth's, from the old Collinsville family that brought starts from Lebanon early in the century. Even when the canes are winter killed back to the ground, the figs usually produce a good crop in September or October, if the summer is hot enough, which it usually is on the Southern Plains. A dry, hot summer also suits the jujube, a tree fruit from Syria, also known as the Chinese date. And it does look a little like a date though its flesh is pale green and has the flavor and consistency of applesauce.

The wild fruits represent a whole world to explore, particularly the orchard cultivars of native varieties. The mulberry trees in the fence rows often bear such heavy crops that branches bend as they would in an ice storm. But the fruit is insipidly sweet, at best an acquired taste. Not so the Illinois Everbearing! Its fruit has all the sprightliness of raspberries. On June mornings I take a bowl out to the tree and pick mulberries for my breakfast, the hens scrambling around my feet for any that fall to the ground.

I can't imagine a life without fruit trees. As a child, candy struck me as boring alongside apples, oranges, and bananas. And a frightening episode in which it was discovered I was highly allergic to mangos left me dejected, for I had liked the mango very much. Even the inedible tree fruits fascinated me—acorns and catalpa pods and Osage oranges—and still do.

The passage I remembered most vividly from The Odyssey wasn't the escape from Cyclops's cave or the encounter with the Sirens and later Scylla and Charybdis, but the description of Alcinous's orchard—"where trees hang their greenery on high, the pear and the pomegranate, the apple with its glossy burden, the sweet fig and the luxuriant olive."

The other evening Berenice scared up two raccoons in the garden and chased them into one of the apple trees. Even after I brought the dog in they continued to chatter, their voices sounding strangely like human conversation, but with a flute-like quality, the "sentences" ending on a drawn-out note of wistful sweetness. There was something so ancient about the sound and at the same time so appropriate to the apple tree that whenever I'm in the orchard it's easy to imagine I'm hearing that music again.

JOURNAL : FALL

(September almost to Christmas)

September 15

Change to Rain and falling moderately, cumulonimbus cloudy becoming stratocumulus mostly cloudy then cumulonimbus partly cloudy then clear with light fog after dark, 69° falling to 63° then 80° F, moist atmosphere, moderate variable winds then moderate S wind then calm, 1.45″ moderate rain from morning thunderstorm. / Froze some green peppers, pressed apples for juice and cider (canning some juice and setting the rest aside to ferment), cut up and froze chicken I cooked in crockpot yesterday. / A new, spring-like season has begun, mellow as spring but built on the solidness of summer instead of winter's fragility.

September 17

Change and falling moderately, clear to cumulus fair, 65° to 91° F, moderate humidity, light S wind. / Inspected beehives, weeded fall garden. / This evening after dusk, while brushing by some branches, a gray treefrog unexpectedly materialized, clinging to my hand. Very adhesive toe pads but not at all sticky or moist— just a smooth, clinging grip. Creamy underside, pale gray overlaid with darker gray markings, each covering about the same degree of area. An intermittent border of tiny black dots separating the gray and cream. About 2″ long.

September 22

Change and steady, clear becoming cumulus fair becoming stratus mostly cloudy, 58° to 75° F, dry air, moderate NNW wind. / Substituted half a day, mixed sugar syrup for feeding bees, made potato soup for dinner. / Big bluestem seedheads (panicles) have turned a rich purple-brown, in a wind the meadows suggesting Homer's wine-dark sea.

September 26

Change and rising slowly, clear, 45° to 78° F, dry air, light northerly breeze. / Shelled popcorn, serviced tractor, sold two gallons bulk honey directly from the tank. / At noon and too high to identify, a noisy band of hawks circled as they drifted south, their cries sharp and bright as sunlight.

September 27

Change and falling slowly, clear, 40° to 75° F, dry atmosphere, light S wind. / Errands in town (feed store, hardware, bank), fixed door at church, replaced several old hive boxes and bottom boards in the apiaries. / Monarchs streaming south, hundreds gathering on blooming tickseed around the pond, along with common sulphurs.

October 2

Change and falling very slowly, clear, 48° to 81° F, dry air, S breeze. / Ironing, cut sesame and spread to dry, finished building a proper calf hut, keeping an eye on Isabel to see if I can forecast her calving. / Most of the big blue-stem ripened to a burnt gold, but still streaked with purple and green and orange with a dusting of powdery blue waxiness.

October 6

Change to Rain and falling slowly, altocumulus partly cloudy becoming cirrus and altocumulus mostly cloudy, 63° to 87° F, humid, moderate to strong S wind, trace of afternoon sprinkles. / Mowed firebreaks through and around my Little Woods, cooked beans, made buttermilk. / High summer lingers in the trees, but autumn is well advanced in the meadow, intricate with the color and texture of flowering, fruiting, and seeding plants.

October 16

Change and rising moderately, stratus fair quickly becoming cloudy, 63° to 69° falling to 50s during early afternoon, moderate humidity, light S breeze then increasing N wind, occasional light mist by nightfall, trace. / A Saudi family came by for honey and milk (unhappy they can't rely on my farm as they would a grocery!), helped friend requeen his two hives. / A colorful autumn: meadows streaked with purple bluestem, fence lines wine-red with sumac, and the electric pole overgrown with poison-ivy a torch of flaming orange. And all day falling leaves pattered like rain on my roof.

November 8

Change to Rain and falling rapidly, cirrus fair becoming stratus mostly cloudy, 42° to 67° F, moderate humidity, strong and gusty S wind. / Made cheese, planted yellow potato onions, dental checkup, long reading evening. / Leafless, the trees whistle and hiss in the wind. Great drifts of lovegrass panicles form on the lee side of buildings and in ditches and against the fence line (enough to belly them, though the prickly seedheads are themselves fragile and nearly weightless).

November 11

Change and falling moderately, stratocumulus cloudy, 45° to 63° F, moderate to strong S wind, mist after nightfall. / Cleared out the garden in preparation for plowing, clamped the turnips. / Between 4:00 and

4:30 an unbroken stream of blackbirds passed overhead, lifting and parting only enough to clear the trees around my place and then dropping down again. Now broad, now narrow—I couldn't estimate the breadth of the stream well enough to get an average count. But in that half hour the total would easily number in the tens of thousands. The flicker of so many wings against that wild sky was such a sight I could do nothing but stand and watch until the last of them disappeared in the southwest.

November 17

Rain and falling slowly, stratocumulus and cirrostratus mostly cloudy becoming clear, 48° to 64° F, moist air, moderate S wind, 1.71" rain—ending early night. / Labeled honey, made chili and crackerbread. / At a time when most other trees are bare-limbed or faded to a dull brown, the white-barked sycamores glow with a mineral richness, gold and copper, the leaves on some clinging until well into winter.

November 29

Change to Rain falling then rising rapidly, stratocumulus cloudy giving way to cumulonimbus mostly cloudy during afternoon then suddenly clear from late afternoon onward, 61° to 68° F, moist air drying rapidly, strong S wind becoming moderate to strong SW wind late in day with some dustiness, .04″ afternoon thundershower. / Gathered kindling in the bottoms (noting there were hardly enough pecans on the ground to warrant picking), cleaned house, pasteurized milk for cheese making tomorrow. / Two dead pear leaves, sewn together and anchored to a pear branch with silk, contained between them a silken bag housing a dead mother spider and dozens of spiderlings—did not open it fully but wedged this spider dwelling in a crotch between two twigs.

December 4

Change and falling slowly, clear, 19° to 46° F, moderate humidity, breeze E becoming S. / Planted more trees in Little Woods, labeled honey, ran errands in town (delivering a dozen quarts of honey to the lumber store). / Pond sleeping under waters so silent and placid that winter winds barely ruffle the surface, unlike summer when there is a sunlit sparkle in the very depths.

December 18

Change to Rain and falling rapidly, clear then cirrus fair, 29° to 67° F, dry atmosphere, moderate SW wind. / Went to Millard Smalygo's for straw, repaired polebarn, pumped the root cellar dry. / These clear days the pond is a hard indigo while the sky is a brilliant cerulean blue.

The prevailing wind ...

The prevailing wind is southerly. Though it doesn't blow as it did during the drought years, the south wind is insistent enough both the trees and the prairie grass lean to the north. Winds from any other quarter are passing events—an easterly bringing rain, a westerly an imminent change, a northerly the passage of a front followed by clearing weather—afterwards the wind veering again to the south.

The southerly wind is what animates the landscape, fluttering the poplar leaves and bringing up Gulf moisture. Without it the region would be arid like the center of most continents. The south wind is a feature of the tropical maritime air mass, and our climate here is not the steppe climate of the Great Plains but temperate humid, or possibly subhumid. Every summer, though, the threat of drought hangs over us, and it's reassuring to have the clouds streaming up from the south—and southwest. In the region of the cirrus clouds the movement is from the south-west. But even the prevailing weatherlies are born as a south wind and deflected to the east by the earth's rotation, the Coriolis Effect.

Clouds are the face of the wind and such a presence against the blue sky that out working I am always aware of them. Like plants and animals, they have personalities and are even classified by the same Linnaean system of genus and species. Clouds are grouped into three families, Cirrus and Cumulus and Stratus. And within each genus there are species, like the Cumulus humilis (fluffy, fair weather clouds) and Cumulus fractus (wind-torn clouds caught up in the violence of a thunderstorm). The thunderstorm itself is of the genus Cumulonimbus capillatus (anvil top), for example, and Cumulus fractus (having no anvil top). I have known them all, watched their birth and their decay into wisps of vapor.

Writing from the pampas of Argentina, Darwin observed that most people rarely look more than fifteen degrees above the horizon, missing most of what occurs there without being aware of it. He may be right, but not about anyone who works out of doors or whose livelihood depends on the sky.

I plan my day according to the sky, laying out the soaker hoses this evening because there's no suggestion of rain in the drift of cloud. And over a season the weather makes all the difference.

To keep better records I built a weather station, a box mounted on a post in the open to house the instruments. And each day I note barometric change, high and low temperatures, humidity, precipitation and its source (.93″ heavy rain from early afternoon thundershower, for example), wind speed, clouds throughout the day, and any notable event.

On a chart I summarize each year's weather in detail and on another the ten-year averages. The decade of the 1980s was unusually wet, with an average annual precipitation of 44.86 inches, but it varied from 56.87 inches in 1985 to 32.65 inches in 1988. A January in 1986 was the driest month (0.00 inches) and a September of the same year the wettest (12.83 inches). The low occurred in 1989 (-14 degrees F), and the hottest year was 1983 (thirty-three days above 100 degrees).

So far the 1990s have been like the 1980s, despite extremes on the whole benign. But always at the back of my mind is a time in the 1970s, before I came to the farm—four months and three weeks without rain, except for one-third of an inch on the fifth of November and a few showers that evaporated without soaking in.

Still, it would be boring to live where there was no drama, no fog "rainbow" of shimmering white light, fading and reappearing … no dust storm in which the sun became a pale lemon sphere with a large sunspot clearly visible … no evening sky overlaid with gold-tinted vapor trails, like pieces of oat straw scattered on a watery surface … no March day, the sky deep blue with brilliant white cumulus clouds, clothes on the line billowing and snapping in the breeze … no sunset with broad streaks radiating from horizon to horizon, shadows of invisible thunderstorms in the distant west … no whirlwind moving across the garden where I was working, filling the air with leaves and blades of corn and insects with glittering wings, lifting them high into the sky.

—A sky in which there are always birds. Even in the iron grip of winter when nothing else is stirring, I have the company of the hawks: the Cooper's hawk, sharp-shinned hawk, and an occasional goshawk (accipiters, long-tailed birds with short wings), the slender marsh hawk (a harrier), the swift and diminutive American kestrel (a falcon), Swainson's hawk and the red-tailed hawk (large birds, both buteos), the latter greeting me when I go out with a plaintive "chirr-chirr-chirr."

In the summer months I never walk alone in the pasture—always with barn swallows flashing past, circling and flashing past again, catching whatever I scare up, insects too small for me to see. The swifts and swallows, flycatchers and nightjars are so much on the wing I associate them with the sky in the same way I do clouds and stars.

The first sound I hear in the morning silence is the shrill chatter of a scissor-tailed flycatcher, the bird hovering against the pale sky. All day purple martins swing through and vanish. And chimney swifts, singly or in tight little bunches, hurl themselves round the house twittering with the mad excitement of it all. And at dusk the silent nighthawks cross and re-cross the pasture in their easy erratic way, twisting and turning.

Cattle egrets, a recent immigrant from Africa, still drift across the evening sky, but not in the same numbers as before. Then, hundreds of the graceful white birds, in scattered V's and streamers, crossed the farm twice a day, moving between the lakes northeast of here and their rookery to the southwest, until it was bulldozed for a new highway. On one summer evening I counted more than three thousand. When the wind was strong from the south they hugged the ground, flying just above the bluestem prairie grass, swooping up to clear a fence or a clump of trees and dropping down again. It was, in the dusk, indescribably beautiful.

Some summer afternoons I drop everything to watch the thunderheads building in the sky, knowing they will likely collapse before sunset, but not always. One August evening around six o'clock I was watching when a rain began to fall. Brilliantly illuminated by the sun, each drop shone white against the dark cloud base. They were clearly visible, several distinct sheets of raindrops, tens of thousands, drifting down in quiet leisure. It seemed a long time before the raindrops I had seen high against the cloud began pattering on the leaves near where I was standing.

The storms that bring summer rain usually develop in the distance, but not always below the horizon. To be present during the entire life of a thunderstorm, to see the whole show, is an unforgettable experience. I remember watching a massive, slow moving storm develop and majestically move across the sky, the tops building to a great height with cirrostratus clouds streaming ahead. The sunset sky was split in two, the west a clear blue, amber along the horizon, while the east, below a perfectly defined thunderstorm anvil, white against the infinite blue, was all rainy darkness emitting a continuous rumbling and flicker of lightning. Long after sundown, towering cumulus clouds drifted across the sky from southwest to northeast, their tops brightly lit, Venus and the moon slipping in and out.

When I'm taking a break from work I sometimes lie on the grass and look at the sky overhead, never without amazement at how filled it is with activity—a spinning top of midges and a pursuing dragonfly, insects too small to identify, the drifting webs of balloon spiders, high-flying birds, distinct layers of clouds topped with wisps of cirrus, and beyond that an infinity of the bluest sky.—"How was it...?" Prince Andrew said, in War and Peace, lying wounded on the field at Austerlitz. "How was it I did not see that lofty sky before?"

High summer is a fifth season on the prairie. From late July through most of September the weather is settled—dry and hot, with a steady south breeze and the singing of insects. In the sunny afternoon, cicadas and grasshoppers, the "zzip-zzip" of the meadow grasshopper and the "crackling" of the short-horned, and later, in the cool of the evening as the first stars are appearing, the angular-winged katydid and snowy tree cricket. Then comes the autumnal equinox.

The best part of night is evening, when the haze of dusk briefly subdues the lights on the horizon, glittering with unnatural intensity, like welding arcs or a carnival midway, chemical blue and cadmium yellow.

In letters from the Indian Ocean, my brother (who was at the time navigation officer on a frigate) describes stars "wheeling across the sky on a clear and perfectly dark night, visible from horizon to horizon."

Once when returning home across the pasture and a hay meadow beyond, from a neighbor's house where I had gone to borrow a book, I navigated by the stars. It was a frosty night and late, a ground fog so obliterated the horizon that none of my familiar landmarks were visible. Even the flashlight I rarely carried, preferring the dark and silence of being out on the fields alone, would have been of little use. And there were no lights visible—none!—except for the stars overhead shining with a brilliance I hadn't seen in a long time. The Pleiades, Perseus, Cassiopia . . . on around the sky until I found the familiar Big Dipper, all seven stars. And steering in the general direction of the constellation, north-by-northwest, I soon found the crossing in the fence and felt the path underfoot. I was met by my cow, Isabel, and by Berenice and the cats, like a man returning at the appointed hour from a long journey.

JOURNAL: THE RANGE FIRE

February 18

Change to Rain and falling moderately, cirrus and altocumulus mostly to partly cloudy becoming mostly cloudy with altocumulus in evening, 49° to 74° F (the high an estimate, the weather station thermometer registering 119° F because of the fire), dry air, S wind strong to very strong.

From 1:30 this afternoon fought a range fire, first on the land south and then in among my own buildings. Isabel I closed in the barn, almost certain it would survive. Until driven away by the heat I hosed down the wood door. The chickens were trapped near the hen house which was sure to catch fire, so a neighbor pitched them over the fence to fend for themselves. They ran to an island of unburned grass and stayed there in a tight little bunch, the fire raging all around. (I would have expected them to scatter in panic—instead they showed uncanny presence of mind.) The dog, the cats (except for Anna), and a rooster I closed in the house. By then a fire truck was on the scene.

In quick succession the hay barn caught fire, then the hen house, calf hut, wood fence posts—and many of my trees, the hardest sight to bear. But all of my figs were saved, most of the fruit trees, and about half my woods. Afterwards I rounded up the chickens and took them over to a neighbor's, then repaired Isabel's pen enough that I could let her out of the barn. A neighbor brought over hay. Anna showed up later in the evening, singed but unhurt.

Complete strangers, men I may never see again, stayed through the long afternoon putting out the last of the burning embers.

Before it reached my place the range fire had run for more than a mile and a half, and it burned on for two miles beyond. The wind and heat waves whipped the flames into vortices similar to whirlwinds, tall and leaning at about a 30° angle—explaining the fire's speed and how it jumped roads in its path.

February 19

Rain and nearly steady, cirrostratus and stratocumulus cloudy becoming nimbostratus cloudy in evening, ground fog with clear above and bright stars, 54° to 59° F, atmosphere becoming saturated, strong S wind all night becoming moderate during day and light breeze evening, moderate rain with thunder beginning in evening ending early night. / Hard to know where to start. Began cleaning up, piling the debris—thinking all the time how I could have made the place less vulnerable to range fires. / Robins, meadowlarks, and redwing blackbirds singing this morning—about the only cheery note in a desolate landscape. Later a crowd of grackles gathered on the burned pasture, moving from one spot to another with a flurry of wings.

February 20

Rain to Change and rising rapidly, nimbostratus mostly cloudy becoming hazy clear then stratus cloudy by nightfall, 46° to 61° F, moist atmosphere, calm then light breeze increasing to light NNW wind by nightfall, .95″ moderate rain ending during previous night. / There have been so many offers I could wind up with more hay than before the fire. Nine came out from church to help with the cleanup, making a great difference. (Two years ago I had lent a hand to one of the couples clearing up after a tornado.) The perfect spring day was such a contrast with the desolate scene that it was almost comic, at least everyone seemed to be laughing and joking as they worked—me included! It made up for yesterday, which was gloomy. And for the day before, which was the very worst! / The moist afternoon sunlight, glittering rain puddles, and crying killdeer lent a suggestion of tidal mud flats to the landscape—an impression heightened by white gulls dotted across a distant burned-over field.

It's the business of a path . . .

It's the business of a path to go unnoticed—until on a dark night your feet tell you something is wrong. You are off the path, lost.

To be "lost" half-way to the barn is an odd sensation. It was on such a dark night I was reminded what a habit this place had become, how familiar the outbuildings and garden and pasture, all linked by paths. A path is, after all, the physical expression of a deeply ingrained habit.

Even the most direct paths have a mind of their own, curving mysteriously, sometimes abruptly and for reasons since forgotten—a fallen branch, a clump of bluestem, the old cattle trail my feet found, something that provoked my curiosity. The willow branch has decayed and the eggs in the meadowlark nest have hatched, and yet the bends in the path are fixed forever.

My path around the head of the pond was laid down in a dry season. And though it is muddy several months out of the year I still persist in going that way, partly out of habit, partly curiosity—to see who else has been along. What looks startlingly like small hand prints are the tracks of a raccoon. And even pressed in the wet soil, the arrowhead shaped hoof print of a deer suggests fleetness. Some are so small I was half ready to believe there was a pygmy species of white-tailed deer, until one day while mowing brush on the adjoining ranch I found a newborn fawn no bigger than one of my cats. Dappled like the wings of a butterfly, it was nearly invisible in the grass. I was so shaken at the thought I might have run over it that I moved the fawn several yards, to a safer place. The small creature bleat-ed like a goat and acted as if it wanted to come with me. I was relieved to see the mother returning as I moved off.

Dusty stretches of path are an even better place to look for tracks. Here an insect as small as a harvester ant will leave its distinctive footprints. And the labored efforts of a dung beetle pushing its burden are easy to read.

—At least until the cow comes along and scuffs the slate clean. Isabel, despite her girth and thousand pounds, walks a narrow path. Little more than the width of a single hoof, her paths are narrower and deeper than I can walk without sometimes getting tripped up, especially in the dark. Since Isabel likes to follow me around when she has nothing else to do, we share the path across the pasture.

Paths link me with a great deal more than the barn and hen house, garden and the pasture beyond. Take an evening in early fall. I had been reading by an open window. As near as I could tell it was a windless night, but the poplars were whispering. Losing interest in the book, I went for a walk along the path to the pond. In the fading dusk the softly glowing sky was utterly transparent, the first stars suspended in the foreground of that infinite depth. Since my feet knew the way, anticipating every twist and turn of the path, I was free to take it all in.

Had this been May I would have seen the Great Arch, an asterism of my own naming, spanning the western sky—Procyon in Canis Minor, Pollux and Castor in Gemini, Beta Auriga and Capella in the constellation Auriga. But tonight the Great Arch lay below the eastern horizon, and in its place lingered the brilliant stars of summer—Vega, Deneb, and Altair.

The evening air is fragrant with sundried cow manure, its pungency more herb-like than animal, the driftwood odors of weathered fence posts and black willows by the pond, and the ragweed's aromatic resins—scents I associate more than any other with the breath of the earth.

A full moon slips above the eastern horizon, silhouetting the wind-blown hawthorns of the upland prairie. And the coyotes break into a yapping and howling, crying out in exultation. It is, briefly, a wild and mysterious scene—until subdued by the soft light of the risen moon pattering like rain on the surface of the pond.

By now it's late, the stellar net of the Pleiades growing brighter in the east-northeast. They rise with the advent of fall. The higher each evening the colder the weather, until the Pleiades reach their zenith in January and then begin the long slide westward into spring.

Walking back to the house, and bed, I remind myself to watch for moonset in the morning over the Osage Hills, more beautiful than any moonrise. The contours of the moon's surface, the melon-orange highlands and blue seas, are then most sharply defined—even as atmospheric effects near the horizon flatten it to the shape of an oval. The moon lingers for a moment, like a distant sunlit peak beyond the shadowy hills, then vanishes. And suddenly it is day.

I am never so conscious of living on a planet and journeying through the universe as when walking on that familiar path which leads from my front porch out past the cow barn, through the gate into the pasture, and down to the pond.

And I rarely come this way without meeting another of the planet's inhabitants—once a horned lizard, which lives with me still.

It was a surprise, finding him in the wet grass at the edge of the path. The horned lizard is typically at home in dryer, more rocky country and has all but disappeared from the area. He was such a quaint fellow, as curious about me as I was about him, and so helplessly out of place where he was, that I took him back to the house and made a home for him from an old hive box with a cover of wire mesh. There, on a mound of gravelly soil with a slab of sandstone on which to sun himself and a chunk of decayed log teeming with ants (his food of choice) he is the image of contentment.

The Texas Horned Toad is one of seven species of horned lizard found in the U.S. ranging from Mexico to the Northern Plains. Of a distinctive flat and broad shape, body about two and a quarter inches in length tapering to a one inch tail, the fellow I ran across is especially colorful—tan and light earth with rosy amber markings, a cream stripe down the ridge of his back, dark brown patches bordered in dusty yellow. Blunt spines cover his back and especially his head, giving the horned toad such a ferocious look that in close-up he often appeared in early science fiction films. The truth is, the spines are all bluff and camouflage; he doesn't even seem to mind being handled.

When I stop by his cage, usually with a piece of decayed log or a cardboard tray on which I've been able to entice some ants with a bit of honeycomb, he watches in anticipation, turning his head and with intelligent eyes taking everything in. He blends so well with his surroundings and catches the ants and other small insects on which he feeds with such a quick flick of the tongue that I'm probably inclined to think him more subtle than he is. But who can say.

Like all reptiles, my horned lizard is unable to regulate his body temperature except by making use of sun and shadow, his shallow water pan and the sandy soil of his cage in which he can burrow. Small as his universe is, he is content. Having shed his old skin and put on a little weight, he will soon be going into hibernation, he and his cage spending the winter in my root cellar. And I'll miss him.

In the country, though, there is never any shortage of company. A beautifully patterned nighthawk spent most of the day perched on the south post of the grape trellis not fifty feet from my door. Out of curiosity I stopped by to check on her from time to time. Usually asleep, but sometimes preening, she left in the late afternoon—to go catching insects on the wing, zigzagging across the pasture in that distinctive way of nighthawks.

Less elegant, but more resourceful in their way, crows sometimes perch on fence posts of the chicken yard and "caw" impatiently for me to let the hens out—knowing it means an easy meal of scratch feed and kitchen scraps. The chickens don't seem to mind. Nor did it disturb them to have a female merlin stopping by on the migration flight south. About the size of a pigeon, this handsome sharp-winged falcon looks a lot like the peregrine. The chickens are usually wary of anything resembling a bird of prey, but must have sensed this small falcon was no threat. I am surprised they were as casual about the barred owl which, just as the chill of early evening was taking hold, came swooping in to perch on a fencepost. A large puffy-headed owl, streaked with brown and cream, the bird was as soft in appearance as the hazy afternoon had been.

The dickcissel, a songbird of the prairie that likes to nest in sumac or small trees, never frequented the farm until I planted a wood. Now the small bird nests here in great numbers, its "dick-chick-chick-chick" heard throughout the summer.

The "Little Woods," as I call it, has already changed the character of Southwind though it is no more than six or seven years in the growing. Fencing off four and a half acres, I began planting well started trees, seventy or eighty a year. Already the larger cast pools of shade on hot afternoons and are the favorite perch of meadowlarks. One redcedar is such a frequent singing post that its central leader is permanently bent and the woody stem polished smooth. The little mulberries have already begun producing fruit. But the best fruit is the day-long singing of the meadowlark and the dickcissel.

The redcedars I mostly dug in the wild, where the cattle would have twisted them out of shape with their rubbing. The white mulberries and black mulberries were "weeds" from my garden. And the seedlings of the common hackberry, sugarberry, and green ash grew rank in the fence rows. The gum bumelia, a small and admirably tough native tree, has been another matter. Not only have seedlings been hard to find, but the mature trees have for several years mysteriously refused to set seed.

For a time I was the caretaker for a small church I attend. Across the street grew an uncommonly beautiful native tree, a soapberry, from which I collected seed. Another native, the catalpa, used to be widely planted as an ornamental because of its showy flowers, large heart-shaped leaves, and long bean-like pods. It's a front yard feature of only small towns now, but I was able to dig all the seedlings I wanted from the church flowerbeds.

Some of the trees for my woods have come from elsewhere. The lacebark elms (Ulmus parvifolia, a true Chinese elm, not to be confused with a dwarf "Chinese," or "Siberian" elm) I ordered as seedlings from the Forestry Division of the Oklahoma Department of Agriculture for twenty-two cents apiece, a rare bargain. The spreading elm frequently grows to eighty feet and lives for centuries.

From ordered seed I have also planted mesquite (as an experiment), the beebee tree (a heavy producer of nectar, for my bees), and the Chinese pistachio. I'm so enthusiastic about this tree (the Pistacia chinensis), which is hardy and pest-free and tolerant of most soils, that I would like to see it widely planted. It's very beautiful in autumn, bearing little red fruits the size of peppercorns which turn blue as winter deepens, and looks as if it belonged here on the prairie.

The Kentucky coffee tree is native and such a handsome tree I went to collect some seed in the wild one afternoon. Because of little wind and the mild, wet weather, not many of the pods had fallen. I threw stick after stick up into the tree, sometimes dislodging a pod or two, more often nothing. Wishing I could have shouted "oo-wong-a-lay-ma," as in the

Bantu story, "The Tale of the Name of the Tree," and had the tree drop all its fruit, I kept throwing my stick. But it was worth a sore arm.

That evening in front of the woodstove, breaking open the pods to get the seeds out, I was rewarded with what looked like a pile of old coins. The Gymnocladus dioica is actually a legume, a member of the pea or bean family, and its large, flat seeds are very distinctive. Early settlers ground them as a coffee substitute.

The stately row of Lombardy poplars which gave the farm such a look of southern France have mostly succumbed to poplar canker and the wood-boring larvae of the long-horned beetle (Cerambycidae). These may be problems endemic to North America since Lombardys are relatively long-lived, even massive trees in Europe and Argentina.

I am replacing the Lombardys with bollena poplars, a columnar form of the white poplar (Populus alba) with silvery leaves and smooth, chalky, pale green bark. It is a beautiful tree, longer lived than the Lombardy, more resistant to fungal and insect damage—and I heartily recommend it.

Still, I will miss the Lombardy's spicy pungency as the new leaves expand in late March. It's a scent distinctive to some members of the Salicaceae, the family which includes willows and poplars. I've recently planted a sapling "Balm of Gilead" poplar (not to be confused with the evergreen native to Jordan)—to fill the gap left by the Lombardys in the fragrances of spring.

As the windbreak near the house and garden I'm planting Osage oranges grown from seed. It's a native tree, drought tolerant and pest free, deriving its name from the fruit, which looks very like a green orange. Also known as "bowdark" from *bois d'arc,* the wood was used by the plains Indians for bows. Later fence posts, railroad ties, paving blocks, and for some reason even chuck wagons were made from its tough, rot resistant wood. But it was as a hedge fence that the thorny tree was widely planted throughout the Midwest in the 1850s. The Osage orange even inspired its own "Johnny Appleseed," a man named John Wright, who spread word of its usefulness. Maclura pomifera is the last survivor in this genus, which as recently as the ice ages numbered many species.

The Russian olive I planted at the corner of the house is not a true olive—but certainly suggests one with its silvery foliage, contorted trunk, and the heavily winy perfume of its abundant star-like blossoms in the spring. I recently planted a variant known as the

Trebizond date, which isn't a true date either—though its one inch red fruits are esteemed in Turkey, Afghanistan, and throughout Central Asia.

From a clump of bristly locust trees growing where there was once a farmhouse I took cuttings which have since become well established. Their branches are covered with flexible bristles, like coarse fur. And in May they flower heavily with long clusters of unscented pink blossoms. The bristly locust is a very distinctive tree-like shrub, stout and contorted, and as hardy as it appears.

I think of it as a family custom to plant a sycamore just after building a house. The one I watched my father plant is now a large tree—as well it should be, since I grew up playing in its shade. I dug mine along the Caney river, from a jutting bankside that was swept away by the next high water.

Winds have a way of shaping the outcome of sycamores growing along upland creeks, molding the trees into heavy-boled fortresses with huge horizontal limbs extending outward to great lengths. Old sycamores, which are often a hollow network of "chimneys," were the original home of chimney swifts. Of all trees I am most attracted to the sycamore, the open architecture of its branches and its fluttering shade suggesting an aspiring and animated spirit.

THE SMALL FARM

Even where the family farm was once a tradition it is no longer. Farmers of all kinds make up just 1.9 percent of the U.S. population. Nearly a third of farm managers and 90 percent of farm workers do not even live on the farms where they work. The farm as a homestead is largely a romantic idea.

And the farm itself is large, the mass producer of a single commodity, a factory. "More economic," they say. And in a consumer society living beyond its means and dependent on purposeless growth to stave off catastrophe, they are doubtless right. Our intensely urban culture would never have been possible without large, highly productive farms.

But such farms, even on a smaller scale, bear no resemblance to a small farm, where the farmer and his family live and raise much of their own food practicing a sustainable form of agriculture. On their farm, even if it is not the sole source of income, they center their existence.

For that is the chief product of a small farm—a way of life.

According to the custom . . .

According to the custom of labeling people by their profession, I am a "Beekeeper." However, since the ton of honey I harvest from fifty-odd hives is my principal source of farm income it would be just as accurate to say the bees kept me.

—We are all kept by bees. Since most of the plants on which we depend for food, shelter, clothing, fuel and beauty would neither flower nor fruit without insect pollinators, like the honeybee, it's surprising how blasé we are. And sometimes worse than blasé.

Because he liked the idea of it, my father once bought three established hives, along with a smoker, bee veil, and gloves. With my mother, we all went out to watch while he unloaded the hives and set them in place. One fell over and we were all pursued by bees, especially my mother in whose flying hair several got entangled, just like in the cartoons. When my father got around to harvesting his honey it was so dark he made mead with it, honey wine, from a recipe in one of Thomas Hardy's stories. In the end the wax moths defeated him—and, as a teenager in need of something defining to do, I became a beekeeper.

Every profession has its mystique, beekeeping more than most. There is the bee itself, by reputation unpredictable and slightly menacing … and the paraphernalia, veil and gloves, bellows smoker and mysterious hive box … and the hocus-pocus, sleight of hand.

The bees recognize me, of course, by my scent and movement, and this is an advantage when working my hives. I wear no beesuit, at most a nylon mesh veil that fits over my hat and ties at the waist. Gloves are clumsy, and anyhow they soon get stiff with propolis, the resinous substance bees gather to use as a sealer. The occasional sting doesn't mean very much, a rising note in the chorus hum of the hive does. Bees are tolerant of meddling, but not of clumsiness.

I can't say that what I feel for the bees is affection, at least not the same as for Isabel and the other farm animals. But I do have immense respect. In the complex society of the bee there is great moral authority. And to be in their company is to feel it. This, as much as the rewards and punishments, is what makes them such good teachers. By the time you are gently brushing bees from a frame of brood comb you will have learned to work smoothly and deliberately. But that is late in the season, and a great deal will have come before.

The season of the bee begins and ends in mid-winter. The solitary bee may venture out on a sunny day, but the colony is clustered in the center of the hive to keep warm, drawing on honey stored in late summer and early fall. I will have seen to it they are amply supplied for the winter, rearranging the filled combs to make them accessible, at the same time leaving space for the clustered bees. The brief lull in activity gives me a chance to paint and repair the hives, see to my equipment and supplies, and review the year.

Like most simple things, the hive with its movable frames is a marvel of ingenuity. Devised in the 1850s by L. L. Langstroth, an American cleric, it replaced centuries of hollow logs and straw skeps and improvised boxes, all of which required virtual destruction of the colony in order to harvest the honey. It was Langstroth who discovered the "bee space," five-sixteenths inch. Anything less is too narrow for the bees to use and quickly filled up with propolis, anything more and the bees build comb across the space, randomly, defeating the purpose of frames and foundation comb. On that bee space depends all of modern beekeeping.

In late February the nectar flow begins from dandelions and silver maples, and the apiary comes to life.

I check the hives again, noting queen vitality and uniting weaker colonies, making sure they have ample stores of pollen and honey. The earliest flowers are so rich in pollen that I rarely need to feed a substitute. But I sometimes feed a sugar solution to stimulate population buildup in lagging colonies and to supplement the stores of bees busy gathering nectar for honey storage.

By March there is so much coming and going of the bees that I remove the mouse guards and entrance reducers. Winter drafts are a thing of the past, and colonies are active enough to keep intruders out. The hives will have shifted slightly with freezing and thawing, and I level them so the bees will build even comb. And I medicate for American foulbrood, preventing the fatal disease of bee larvae before it can get a start. The treatment is innocuous and doesn't get into the summer honey.

Since healthy bees tend to stay healthy, I try to ward off disease through maintaining strong hives. That means keeping a sharp eye throughout the year—for American foulbrood (especially in swarms I've been asked to capture), nosema infection (in spring), chalkbrood (a genetic predisposition corrected by requeening), tracheal mites (a wintering problem treated as if it were a case of croup, with menthol fumes), and Varroa mites (of which I'm wary, though my hives have never had them).

Neither Varroa nor tracheal mites are native parasites of the European honeybee (Apis mellifera), but in recent years have spread from Southeast Asia. There the mites are little more than a pest of the Asian honeybee, with its shorter brood cycle—but here, among the Italian bees of my apiary, they could be devastating.

On the other hand, the threat posed by the introduction of the Africanized bee has been wildly exaggerated. Its aggressiveness appears to be fading as it spreads. In any case, it doesn't store enough honey to survive the winters in most of North America.

Even the wax moth, the old curse of hobby beekeepers, is rarely a problem if hives are strong and the drawn extracting comb carefully stored. —Still, beekeeping isn't simple, but then it never was. And in its difficulty lies opportunity for the small producer who alone can afford to give proper attention.

In March I reverse the brood chambers, something I'll repeat later in the spring. The queen tends to move upward in her egg laying, the bees gradually vacating the lower brood chambers and crowding into the upper ones. This contributes to swarming in April or May, the old queen leaving the hive with most of the workers to establish a new colony, sharply reducing the hive's productivity. Some swarming will be unavoidable—usually on a still, warm day between ten o'clock in the morning and two o'clock in the afternoon, the bees clustering on a lower branch of a nearby tree for several hours before moving on. I can almost anticipate when this is most likely to occur and will try to capture and hive the bees.

Sometimes I get anxious calls from people with swarms in their yards. I'd rather not take on bees from an unknown source, but sometimes do, shaking the cluster into a twenty gallon pail with a lid, both pefo-rated for ventilation. It sounds daring, and a swarm of several thousand restless bees can look terrifying. In fact, bees are least likely to sting on that occasion. They are at the moment homeless, and bees rarely sting except for the best of reasons, in defense of the hive.

The colony is everything. While it may acquire something of a personality, at least a disposition, the individual bee has none. I sometimes think of it as an exquisite little machine, precise and jewel-like. If a worker bee, one of the fifty thousand or so undeveloped females that do the work of the colony, is dehydrated or weak from hunger it will be fed by the others. But when one is injured it bleeds a fluid, like a leaky transmission, and is discarded from the hive as if it were a piece of broken equipment.

The workers as well as the drones, the several hundred males in the colony, are all offspring of the queen. Except when overwintering, they usually live for only a few weeks, and their role in the hive is limited. The queen, who directs the work of the hive through a set of scent factors (pheromones), may live for several years, but seldom does.

I can almost always tell when a colony is queenless as soon as I open the hive. There is a restlessness and a hollow note in the buzzing. When a queen begins to weaken or dies, workers raise a new one by placing a newly hatched larva in a special queen cell and feeding the larva a diet of royal jelly. Occasionally they aren't able to replace the queen, and the scene of despair in such a hive is oddly disturbing. The entrance is left unguarded, the hive dirty and diseased, the bees apathetic, resigned to the colony's death.

To prevent that, I requeen half my hives annually, with new queens from outside breeders, occasionally adding package bees as well. Until now I've been relying on Starline queens, a hybrid strain of Italian, one of the European honeybees. The Caucasian bee is reputed to be of a milder temperament, but I get along quite well with the Italian. And they are the better for mid- and south-temperate latitudes. The Buckfast queen shows remarkable resistance to tracheal mites, and I have been introducing a few into my apiary. It was a Benedictine monk, Brother Adam, of Buckfast Abbey in England who selectively bred the European honeybee for more than seventy years (he is ninety-four!) to develop the improved strain.

The requeening method I use is, again, labor intensive—but almost invariably successful. The same can be said for the method I use for adding bees to an established hive. Bees are notoriously xenophobic, and introducing strangers takes a certain amount of tact.

By early April the strongest colonies may have to be divided to discourage swarming and frames of brood transferred to the weaker colonies to bring them up to strength. And I check the laying pattern of the queen, and the queen herself if I can find her, which usually isn't hard. I can tell where she is from the behavior of the bees, their agitated buzzing when I remove the frame containing the queen and her court of attendant workers.

I rarely open a hive without seeing something that needs to be done, like replacing the old and blackened brood comb. Each generation of larva leaves behind a skin during pupation. Though workers polish out each vacated cell with a coat of beeswax, the cell darkens and narrows, resulting in smaller workers and a declining honey harvest. The object of all this shuffling is to build strong colonies and colonies of equal size and productivity, which makes for easier management.

By May the nectar flow is at its height, the bees' nervous industriousness pervading the whole farm. A bee can carry as much as eighty percent of her own weight in nectar and pollen and may visit from fifty to a thousand blossoms, some as far away as two or three miles. In a day the bees from a single hive might visit a million and a half flowers and fly a total of thirteen thousand miles. In the evening the bees cluster in drifts on the outside of the hives to cool and rest. And the air is fragrant with the heady perfume of ripening honey.

It would be impossible not to be busy at such a time, and I am busy—installing queen excluders at the beginning of the nectar flow, to prevent the queen laying eggs in the honey supers, and adding more supers as needed, setting them a half inch forward to provide additional hive entrances.

In July, as soon as the main nectar flow has eased to a trickle, I harvest the honey—first driving the bees down out of the honey section with a pungent bee repellent sprinkled on a fume board placed over the frames of the top super. It takes less than five minutes to drive the bees down, but longer to free up and move the supers, the boxes containing the frames of filled comb. They are stuck securely together with propolis and may weigh seventy-five pounds, depending on their depth. Hot work on a summer afternoon (which is when it must be done) but infinitely satisfying.

On request I have sometimes collected pollen using a simple device called a pollen trap, which loosens it from the bees as they enter the hive. This means more workers have to spend their time gathering pollen, a principal food of the colony, so it reduces the amount of surplus honey. I'd rather not bother, preferring to concentrate on honey production. Still, I miss the chance to study the colors of the pollen—pale green, pink, orange, magenta, gray, almost black, cherry red—and may one day devote a hive or two to its production.

Bottling the honey, because I've known the whole of the story, is work filled with a kind of suspense. I know it was a long winter and a wet spring; but I also know it was a summer of such abundant blossom that there will be a record harvest. And from that first sampled taste of the honey until the last of it flows into the last quart jar, I'll savor my success—and that of the bees.

I've prepared jars beforehand, and the honey house—blacking out all of the windows except one, which has an improvised bee escape, and filling any chinks. After the stray bees are out of the supers I can uncover the windows. The bees and I play a game for the first few days, the robbers looking for any crack through which they can invade the honey house and me blocking it up.

For uncapping the honeycomb I've been using an uncapping fork, an electrically heated knife taking off too much of the comb. Next year I may try a honey punch, a simple device that rolls over the surface of the comb, breaking a hole in each capping. It sounds like it would be faster, neater, and more efficient. Almost too easy, but I'll try it. Since I work under warm conditions the honey should run out well.

The uncapped frames are spun in a 20-frame radial extractor, centrifugal force draining the honey away from the comb in twenty or thirty minutes. While the extractor is running I uncap more combs and hang them in my frame cart, the cappings collecting in a tank. Once they've drained, I'll melt the caps in a homemade solar melter, using the wax for making candles and for mixing with paraffin to coat my cheeses.

The supers of extracted comb are returned to the hives for clean up by the bees, then stored in the honey house for next year. When the bees don't have to draw out comb, constructing those hexagonal cells of wax, they can spend more of their time collecting honey, which pleases us both.

The extracted honey, twenty-two hundred pounds of it, is transferred by bucket to my strainer and storage tank—and the jars, mostly quart size, filled from there. To some I add sections of cut comb. It's tedious and time-consuming work. But I can appreciate why customers want to have a sample of the honeycomb—which is, in its simplicity and utility, as perfect a technical achievement as could be imagined.

—But no more than the honey. To make it, nectar is converted by the addition of the enzyme invertase, which breaks down sucrose into a variety of simple sugars, many of them obscure and in minute quantities. Simple sugars keep better and take up less storage volume. During a ripening process in the comb, excess moisture is reduced even further. In addition to water and sugars, honey is composed of plant acids (the source of its subtle flavor), minerals, enzymes, vitamins, dextrins, colloids, and inhibine, a natural antibiotic.

As it turns out this is a vintage year, the honey pale gold with a summery bouquet—brilliantly clear and yet, when a jar is held to the light, fathoms deep.

August and early fall is a quiet time. I have sometimes harvested a little of the late honey, from ironweed and eupatorium and goldenrod, but ordinarily leave it for the bees' winter stores.

A rule of good beekeeping is to take winter losses in the fall by never trying to winter over weak colonies.

Unite the weaker hives, replace any aging queens, medicate for the winter diseases, be sure there are ample stores of honey, arranging these to accommodate the winter cluster of bees, and reduce the entrance size, adding mouse guards as cold weather approaches.

On a day in January when snow is drifted around the silent hives, I am struck by the sharp contrast between winter and summer—when the bees, their wings glistening in the sun, are crisscrossing in such numbers the air appears to be strung with gold wires.

THE SOLITARY LIFE

Following the feverish activity of summer, building my house and barn, but before weather would permit getting into the garden and before there were animals, I experienced a wintry interlude of loneliness. School friends had scattered and I wasn't very good at striking up new acquaintances. Married, things might have been different, but I wasn't.

With the coming of spring there was no time for loneliness, even though running a small farm remains a solitary business. A honk from the passing school bus counts for social life when you live in the country, sometimes passing whole days without talking to anyone. But I am not lonely, a word I would never think to use now.

It was the farm itself that turned out to be the best company—a full schedule of work, a Jersey cow scuffing along at my heels, the sun and warm wind. Sometimes I wonder if it isn't for this kind of life that what we call "loneliness" is really crying out for. After all, this is the world in which Homo sapiens evolved—and not the manufactured environment and the video reality. That it is filled with people, this urban world, does little to make it less lonely.

Increasingly the solitary life (even at the risk of passing loneliness) is itself an adventure. Increasingly I am attracted to solitary places and solitary lives. Every morning I look first toward the distant Osage Hills, glowing pink with the light of sunrise, cheered by the thought that the last large tract of tallgrass prairie is so close—and so empty. And I am attracted by the experiences of men who built small boats and sailed them around the world—Joshua Slocum, Harry Pidgeon, Jacques-Yves le Toumelin.

—Like them, I was a free man and my own master and looking for adventure when I came to the farm. My prairie horizon is almost as wide as theirs, and I am almost as dependent on the wind and weather. Even the journal in which I write up each day's events is like a ship's log. And what am I doing on the farm if not outfitting and provisioning for an independent voyage of unknown duration.

As milk cows go . . .

As milk cows go, the Jersey is small. Having been born on the farm, fed from a bottle, brushed every day and made over, Isabel is inclined to be as frolicsome as Berenice, my terrier. But when at milking time I lean my head against her side, she steadies. Jerseys are known for their calm disposition, for the richness of their milk, and for their distinctive fawn color.

In winter Isabel's coat is woolly and a pale reddish brown; in summer it lays very flat and is much paler, a sun bleached tan with darker shading on the forelegs, cheek, and tail. The long, coarse hairs on the "switch" of her tail are almost black. The first cow flies mark the coming of spring, in March. And a sudden slap from Isabel's tail can leave such a stinging impression that I loosely tie it with a sash to her hock while milking.

Despite her tail, despite everything, milking is a pleasant time for us both. Except when drying her off,

I milk twice a day—at 6:45 in the morning and again at 5:30 in the afternoon. If I'm even a few minutes late, which is rare, she bellows impatiently. The dairy ration I feed her, smelling sweetly of corn and molasses, has something to do with it, but no more than the routine itself, in which she takes an engrossing interest.

I wash her teats with a weak bleach solution, put a little Vaseline on my fingers, squirt a few streams of milk on the straw bedding to clear any dirt, and then begin to milk. The "letting down" of her milk begins when Isabel hears me coming. And by the time the milk is roiling in the bucket, frothing like sea foam, milking has become quite easy.

Jerseys have a reputation for being difficult milkers because of their small teats, smaller still since the introduction of milking machines. Certainly it's harder than milking a goat, in which you can use the whole hand. A cow is milked with the fingers only, a stripping motion that takes some getting used to. I had never milked a cow until the one I had bought stood impatiently in the stanchion. But now it's second nature.

I sometimes close my eyes and slip into a reverie, the muscles of my hand and arm working automatically, my head resting against her flank. Forewarned by a twitch, I even tilt the bucket aside from a shifting hind foot without really waking. Around the handle "milk stone" has formed, a lime-like deposit that builds up in minute increments over a period of years, so smooth and hard it's part of the bucket.

After milking I dip each teat in a mild antiseptic to reduce the chance of bacterial infection and turn Isabel out of the stall with a pat. Moving across to the hay rack, she tucks in with enthusiasm, putting away a great quantity of coarse, dry hay where fermentation begins in the rumen, accompanied by the sounds I've been hearing as I milked, like the plumbing in an apartment building. Later on she'll ruminate, burping up a fist-sized ball of softened mash for further chewing. It's pleasant being near Isabel while she is chewing her cud. That and the clucking hens are the most soothing sounds on the farm. I sometimes think of all the anxious, troubled people who would find a peace in just listening.

I restrain Isabel's milk production by feeding no more than eight or ten pounds of dairy ration. It is better for her. Even so, she gives seven or eight thousand pounds, around nine hundred gallons of very rich milk every year, four gallons a day. Most of it I sell from the farm as raw, or unpasteurized milk, with the naturally occurring lactic acid bacteria alive. It's more digestible and nutritious. Since Isabel is vaccinated and isolated from other cattle and her health watched over so closely, her unpasteurized milk is quite safe. And the customers who come, some from a considerable distance, are also glad to have any extra eggs and vegetables from the garden.

It's apparently a custom in some parts of the world to buy directly from the farm, and I've become acquainted with an immigrant family from Russia and another from Saudi Arabia. It was a novelty to glance up from my milking the other day and see a veiled woman standing in the open doorway of the barn.

For some reason I haven't specifically worked out the economics of owning a cow. There is the milk for my own use, of course—fresh milk, cheese, butter, and cultured milk. And from the sale of milk I make enough to pay for the hay and feed and most expenses. The registered calf she bears each year brings in another two hundred dollars or so for a heifer. Manure from the barn goes to the garden and any unsold milk to the chickens. But for me there are values in having a cow that wouldn't appear in any column of figures.

Living with a milk cow, winning her affection and respect is in itself something worth doing. Isabel is a large animal with sharp hooves, and we both know it is with her permission I am there at her side with my milk pail. Of a calm and peaceable disposition herself, a cow expects the same from you. If you haven't learned your manners, a cow will teach them to you. In fact, there are a great many qualities in a cow's nature that in a man or woman we would equate with character.

This is not to say a cow is always placid. Isabel can be frighteningly playful, so overflowing in high spirits she sometimes frisks around me, circling and kicking up her heels, or stampedes down the path after me, hooves thundering. And she takes an impish delight in catching me unaware and nudging me with her battering ram of a head or bellowing in my ear or, more often, wrapping her long coarse textured tongue around a sleeve or pant leg and giving a mighty tug. Isabel likes to be brushed, and has been brushed every day since shortly after she was born.

That a cow could be such a good companion, so lively and responsive, might seem absurd to an outsider who sees only the flies and the bother. I was wrong, though, in assuming everyone responded as I did to the sight of a Jersey cow grazing knee deep in a grassy meadow on a summer evening ...

Sophia, Isabel's mother, had already been cut from a dairy herd as too low-producing a cow to be economic and was destined for the slaughterhouse when I bought her. She was an ideal milker for my purposes and made a dairyman of me. But she was getting older, and I decided to keep the next calf, if it was a heifer. And contrary to the way such things usually work out, it was, a duplicate of Sophia herself.

Dairy cows are bred three months after calving, by artificial insemination. Jersey bulls are so dangerous it wouldn't do to have one around. The cow lets me know when she is in heat, and a friend who until recently ran a Jersey dairy comes by in his truck, bringing the semen of various registered Jersey bulls stored in a cannister of liquid nitrogen. When it's opened, clouds of vaporizing nitrogen pour out, hissing and spitting, and from deep within the cannister come ominous knockings. In the dim evening light, the breeder is like a wizard, his face veiled in clouds of nitrogen as he reads by flashlight the tiny strings of figures printed on the slender tubes containing the semen—and with a triumphant gesture pulls out the proper suitor for my cow.

As the calving date approaches and the cow's udder becomes tight, I confine her to the holding pen. There has never been any complication with Sophia or Isabel. Heifers always come early, bulls always late. The calf, its circulation stimulated by the mother's licking, is on its feet almost immediately and soon nursing. The colostrum, the important "first milk," is thick, rich, antibiotic-laden, and looks very much like canned milk.

Isabel was born at 3:00 in the morning and over a week early, so early I hadn't yet confined Sophia to the holding pen. Awakened by a single bellow from the pasture, I went down to where they were and carried the calf up to the stall, the mother following anxiously along behind. A Jersey calf, with its fawn coat and large doe eyes, surpasses anything the most sentimental Disney illustrator could invent . . . and is at the same time drolly comical, wrinkling its nose as if to say it found you odd and somewhat distasteful.

For the first several weeks Isabel was unable to stand properly, her front hooves turning under at the first joint. At the vet's suggestion I made padded splints from two-inch plastic pipe, sawing and filing until I had two ten-inch lengths cupped to fit her legs. From the first, Isabel possessed a distinctive nature, and in the course of fretting over her splints a strong friendship developed between us, one that deepened when I weaned her over to the bottle.

It's always best if after a couple of days the calf is separated from the mother, easier for both if done before a strong bond is formed. I make a point of giving the calf a lot of personal attention during this time. Before selling a calf, always to someone I know will treat the animal well, I bottle feed it through the period when it is most vulnerable to scouring. Some of the heifers will become family milk cows, and from time to time I hear news of them. The bull calves will go for beef, most raised by neighbors on back lots, far removed from the terrors of the feed lot and cattle truck and packing plant.

Training a calf to bottle feed is typical of a hundred tricks the small farmer must learn, and does as a matter of course. Since the calf is still unsteady, it's best to back the animal against something and then straddle it. Slip your fingers into the calf's mouth, and, once it's nursing, ease the nipple of the bottle in. Hold the muzzle and nipple steady with one hand, and with the other hold the bottle no higher than the calf's back so that it won't choke. When introducing the first solid food, slip a few grains of calf starter in just as it's finishing with the last of the milk in the bottle. It's oddly satisfying, bottle feeding a calf.

In fact, every aspect of feeding animals is something I enjoy, even watching a cow graze. A cow is content with a variety of pasture grasses, some weeds, tree leaves, and legumes. Winter forage consists of vetch, sweetclover, fescue, and rescue grass. (The first to grow rank in early spring, rescue grass has saved many a settler's herd on the brink of starvation.) These days cows are fed hay, both prairie hay and alfalfa, during the winter and whenever grazing is sparse. The alfalfa I feed, thirty-five bales each year, is grown on bottom land along the nearby Caney River. And the sixty bales of prairie hay, some for the garden, comes from down the road.

When on a winter day I break open a bale of prairie hay, the fragrance of summer wafts up around me—pungent ragweed and the dusty green scent of big bluestem and Indiangrass. And there are the grasses themselves, still in shades of green, and the flowers—lavender blazing stars and the golden ashy sunflower.

Of all the colors, that of the bedding straw is in its way most striking. Each year for the cow and hen house and garden I buy fifty bales of wheat straw, which is a sandy gold with a satiny sheen, or oat straw when I can get it, a metallic brass-gold. In the fairy story, remember, it was straw that was spun into gold.

Even doctoring a cow became an interest once I had a little experience. Experience came quickly—a case of bloat. It's happened only two or three times, but whenever we have a sudden lush growth of grass and legumes in the spring I watch for the first symptoms of frothy bloat. When masses of small air bubbles form in the rumen and the cow is unable to belch, her whole digestive system is stopped up and she is in considerable discomfort. It's cured with a defoaming agent, specifically with a common household detergent, "Tide." Mixed with honey or molasses and coaxed down the cow by whatever means, this usually does the trick. The hardest thing was finding a store open that time of night.

While still my source of advice and medication, the vet rarely calls at the farm anymore. Vaccinating, worming, hoof trimming, and the routine doctoring I can take care of.

When Isabel was several months old I was able to reunite the calf and cow. By then Sophia had begun to dry off, my milking down to once a day. She was an older cow when I bought her, and I had already decided not to breed Sophia again—but to keep her on the place for as long as she lived, an experience almost no domestic cow ever knows. There was plenty of pasture, and it was a beautiful sight, the two of them out grazing together, especially in the evening.

—One evening in particular. A passing shower had left the grass wet, and at sunset the sky cleared except for a few broken clouds in the west. The first stars had just begun to appear when a car passed, turned around in the road, came back, stopped—and a man with a rifle got out, stepped to the fence, and fired three shots, killing Sophia.

Since I couldn't identify the car, there was nothing the sheriff could do. It wasn't an accident, but neither was it personal—just something that happened.

My time at Southwind Farm had lulled me into thinking this was an orderly universe, essentially benign and tranquil. Droughts, high winds, late frosts, blights, insect pests, sickness in the animals ... are natural occurrences, none taking me entirely by surprise. There was always something I could do about it. Here there was nothing—except for emigrating to New Zealand, which I briefly considered, or bringing Isabel up closer to the house, which I did, fencing off a part of the pasture.

The spell of the place, though, was broken. Permanently, it seemed.—But I hadn't appreciated what a daily round of chores and the preoccupying interests of the farm could do to mend things. I never think of it now, that summer evening, except as something remote and alien.

STAYING PUT

Coming to the farm was itself the end of a journey. What is "home" if not that place? And the best evidence of having arrived is that I do not feel compelled to always be leaving.

—Which is just as well. It is impossible for a small farmer to be away for more than a few hours without making the most extensive arrangements with neighbors, and then worrying. It sounds easy enough—the instructions you've left for letting the stock out in the morning and putting them up at night, for gathering the eggs and milking the cow—but there are subtleties no one could have dreamed of who has not been a small farmer.

And besides, who can milk a cow these days? And what will Isabel have to say about it? That is why people who come to the country seldom leave home. In a nation where every year one-fifth of the population changes its address and twice a day everyone goes somewhere, this seems inconceivable. But it is not inconceivable if you live on a small farm, and it's not to be regretted.

Once a year (while Isabel is dry) I go to visit my brother and his wife, staying away for as long as four days. We visit botanical gardens and eat at Russian restaurants, and I have a very good time. But I am always anxious to get back to the farm—where things are happening every day that will never happen again and I am missing them. Away from home it's the little things I miss most, like silence.

I once heard an experienced traveler say that the more he saw of the world the less he knew of it. That's how I feel about my twenty acres, where I'm just beginning to learn my way around. And since we are all travelers, traveling the Milky Way together—migrating birds crying down from overhead, I leaning on my hoe—it doesn't really matter which of us is passing through and which is staying home.

A field of ripening wheat . . .

A field of ripening wheat, even a very small one, is sublime. Nothing catches the light and wind like hard red winter wheat yellowing in the sun. It's enough to make me wish I had planted more, and next fall I will. But never so much that I can't harvest it by hand—scything, flailing, winnowing the grain, and at last milling it into flour.

The raising of small grains is something apart from gardening, more rooted in myth and antiquity. For ten thousand years wheat and its kind have been a staple. Nothing stores as well or can be served up in so many different forms as grains—unless it be some of the legumes, which are so complementary I think of them in much the same way. But no other crop is such a feature of the landscape as a wheat field.

Of small grains I raise the following: hard red winter wheat, grain amaranth, sesame, flour corn (Alamo-Navajo Blue), popcorn (South American

Giant), and canola (Westar). And though always trying something new, I plant at least enough of these legumes for my own use: soup peas (Maestro), yellow soybeans (Maple Arrow), crowders (Running Conch), mung beans for sprouts, and multi-purpose dry beans (Dwarf, Horticultural, Maine Yellow Eye, Black Turtle, and Navy). Beans, corn, and amaranth were the staple crops of the pre-Columbian New World. And for many of the same reasons they have become mine.

Amaranth uses the C4 carbon-fixation pathway, a type of photosynthesis especially efficient at high temperatures, in bright sunlight, and under dry conditions. The more commonly used C3 pathway requires more water in photosynthesis, water that is not always available in late summer on the prairie. And with its flamboyant rosy plumes, amaranth is worth growing for its bloom alone, quite apart from the high nutritional value of its seed.

I sometimes wonder why I bother with wheat, especially since a bushel of freshly harvested grain of high quality can be bought at the feed store for something like $5.00. But when October rolls around I always do plant a small field of wheat—digging out the dock and other perennial weeds, cultivating the soil twice with a tiller, and then broadcasting the seed with a cyclone sower. To get even coverage I broadcast half the seed over the entire field walking north and south, and half walking east and west, scattering the seed at the rate of about six and a half pounds (four and three-quarter quarts) per 2000 square feet.

> *Sow four grains in a row*
> *One for the pigeon, one for the crow,*
> *One to rot, and one to grow.*
>
> —*rhyme of the sower*

I've learned to judge the evenness of the planting as much by the steady whirring sound of the hand-cranked cyclone seeder as by watching to see where the grains of wheat are falling. Afterward I cultivate the sown ground lightly with a tiller to cover the seed.

Since wheat is sown in the fall, and in February is the only green in a dun colored landscape, it has already established a good root system before the spring and following summer demands rank growth for grain production. By late June the yellow wheat is ready to scythe, the time to cut a matter of some deliberation. I wait until the rich gold has bleached to a sandy color, but not become dull or grayish. When chewed, a kernel of wheat should be firm, not too hard or the heads of grain will shatter too easily when being scythed.

Scything is a subject all by itself, infinitely subtle. It looks so easy, and is when well done, smooth and rhythmic. The object is not to hack at the wheat, propelling the blade with arms and shoulders. Scything is a twisting motion from right to left and back again, the whole torso coiling like a spring, storing energy at either end of the swing.

Each pass with the scythe leaves a crescent shaped swath of cut wheat. My swaths are not very wide. Instead, I try to keep the heel of my blade down, especially at the beginning of the stroke, concentrating on an even and cleanly cut swath. The sound is like a sigh, and when the scythe takes possession, the indolent swinging motion strangely satisfying.

The scythe I use has a bent snath of tubular aluminum, but I intend to replace it with a straight ash one, which I think would be less unwieldy. The twenty-seven inch blade was too long for my style of scything, so I cut four inches off the tip. Hammered blades are supposed to be superior, but my stamped blade from Austria is easy to sharpen with an electric drill mounted grindstone. Hammered blades require peening with a hammer on an anvil. Final sharpening is done with a whetstone, about every twenty minutes when scything wheat. While cultivating the garden I turned up a whetstone the Dutchman must have used to sharpen his tools, maybe even a scythe.

After scything a swath or two, I gather and tie the cut wheat, later leaning the bundles upright against a fence for further drying. When the wheat is thoroughly cured so the heads shatter easily, it's ready to be flailed. Breaking a bundle open, I spread it on an old bed sheet, folding it over to keep the grain from being scattered. After each hundred strokes with the flail I stir the straw and flail some more, until the grain is separated from the head. The work is rhythmic and just strenuous enough to be relaxing. And I enjoy very much flailing wheat as the day passes around me.

My two-piece flail, patterned after one used since the fourth century, has a five and a half foot handle and an eighteen inch swingle, two and a half inches in diameter. The swingle (traditionally a "cubit," the length of the forearm from elbow to the tip of the middle finger) is attached to the handle by lightweight rope looped through a swivel of heavy gauge wire, free to rotate around the end of the handle in a notch. My first flail I cut from an ash sapling, the one I use now from oak.

After flailing all the bundles, I give the straw to the chickens so they can glean what grain remains. The harvested wheat I winnow, pouring the grain from bucket to bucket, the wind carrying away the chaff and bits of stem. —Some of the wheat, too, when a gust comes along, which is why I prefer to winnow in front of a fan. I usually wash the grain to float off the last of the chaff, and then cure the grain in the sun for a couple of days before storing it away in my pantry.

With a little diatomaceous earth mixed in to deal with any insects that might hatch, the wheat will keep almost indefinitely. Because of the quantity, it isn't practical to heat treat my wheat and blue corn to kill insect eggs as I customarily do amaranth, canola, popcorn, and sesame seed (oven heating to 150° F for 20 or 30 minutes). Diatomaceous earth, the fossil shells of microscopic algae so sharp edged for insect larvae, is harmless to humans when used for this purpose. But I winnow the grain again before milling it into flour, usually on my doorstep since the lightest breeze is enough to blow the powdery diatomaceous earth away. The grain that remains, even the wheat I've bought at the feed store, is much cleaner than most used in the food industry (which has been stored for months, even years, and where the allowable amount of insect parts and rodent droppings is actually spelled out).

Blue corn would be my staple source of flour if it could be made into yeast bread—and if the sight of ripening wheat were not so irresistible. Harvesting corn is so much easier. I have learned to break the ear right out of the husk, leaving the husk still attached to the stalk. If the ears are fully mature and the kernels well dried, even husking is quite easy. Some use a peg to rip the husk from the ear. But I prefer to hold the stem just below the base of the ear with my left hand, strip enough of the husk free to grasp the ear inside with my right hand, and snap the ear free of the stem.

The kernels are stripped from the husk with a hand cranked corn sheller. There is so little chaff, winnowing is easy. Before storing the corn, I let it dry in the sun. Corn of any color is beautiful, but I'm especially attracted to Indian corn—the kernels of my Navajo Blue shades of steel gray and deep purple, sometimes marbled, and with a hard enamel gloss.

Popcorn, which also comes on in July, is processed the same way. Unappetizing as it may sound, a bowl of popcorn with milk and honey is a treat. Millet, a grain similar to corn but more tolerant of heat and drought, is a staple in Africa. The first autumn in my house I bought a 50 pound sack for $3.00, adding it to the wheat in my hot cereal.

Amaranth—cultivated more intensively than wheat or blue corn, and requiring deep, fertile soil for quick growth—is more of a garden crop. The seed grains are so tiny plants have to be thinned to prevent crowding. I usually do this twice, early in the season, thinning to 4 inches apart in rows 24 inches apart.

Amaranth is ready to harvest in July, when the seeds can be easily rubbed out of the heads and are firm when chewed. I cut the heads with pruning snips and spread them on sheets in the sun to dry, turning them at least twice and drying until the heads bleach out and are quite crumbly. As with wheat, I flail the heads in folded sheets on a hard, dry threshing floor, turning once. Since there is a lot of chaff and the grain is small, I sift—using a sifter box made from a shallow depth beehive super and 1⁄16 inch mesh screen (the familiar window screen). The grain and tiny chaff, mostly grain hulls, fall through as I vigorously stir the grain. The larger doesn't, and is discarded. Next I winnow in front of a fan on low speed.

The clean seed that remains may seem little to show for the effort. But amaranth flour is highly nutritious and lends such a delicious nutty flavor to breads, cereals, cakes, and cookies that I'm surprised it isn't more widely used. Once it was, along with spelt, quinoa, and teff—other ancient grains overdue for a revival. Since harvest does not coincide with the beekeeper's busiest time, I intend to rely more heavily on amaranth in future.

Sesame, which is ready to harvest in September, I think of as a kind of Old World equivalent of amaranth. It's a staple in Turkey and the Middle East. I harvest the bushy plants with pruning snips and spread them on old sheets to dry until the thick green stems have turned brown. Then I flail as with other grains and sift, using a sifter box with 1/8 inch hardware cloth, then winnow.

Canola (rapeseed, a member of the mustard family) I raise for sprouting. Harvested in July, the sprouted seeds are a ready source of greens throughout the winter, supplementing those from my cold frames. If I grew my own oilseed crop, this would be it.

Throughout the summer and fall I'm harvesting a succession of legumes—soup peas in June, dry beans in July and August, crowders in August, and soybeans in September. When the pods are dry and most of the leaves shed, I cut the plants off at the base with pruning snips and spread them on old sheets to dry for a week, more or less.

As with grain and seed crops, I flail the dry bean plants between the folds of a sheet spread on the threshing floor, turning at least once. I next dump the beans and plant debris into a large bucket, and sifting through it with my hands, throw out the larger pieces of chaff. After winnowing the beans, pouring them slowly from bucket to bucket in front of a fan (first on medium and then on high), and floating off the last of the debris and bad beans, they are quite clean. I sun dry them for a short time and then heat treat them in an oven (150° F for 20 minutes) before storing the beans away, by now glossy and beautiful.

One of these days I'll build an airy shed to make drying the various grains and beans easier—so I won't have to move them to shelter every time rain threatens.

Hand grinding the grain into flour is real work. I use both hands to crank the mill. My heavily constructed flour mill (a "Corona," made in Colombia) is bolted to a pantry shelf made of two-by-twelves. So it will be fresh, I prefer to grind only enough flour to last a week or two, but even that takes a little time.

—And to pass the time I sometimes look out the pantry window or watch the seeds churn in the hopper or just close my eyes. Once, having fallen into a kind of sleepy reverie, I was surprised to glance down and see my left elbow automatically weaving from side to side as I cranked—to avoid hitting the thumbscrew on the crank handle. I must have learned to avoid that jutting thumbscrew years ago, and then forgotten, the subconscious mind looking after my elbow.

By setting the distance between the steel burrs, I can grind various grains and seeds, peas and beans for any number of uses. For whole-wheat flour I grind twice, a cup of grain yielding a cup and a half of flour. I also grind twice for amaranth flour and for cornmeal, adjusting the burrs to run closer for the second grinding. Soybeans and split peas I coarse grind only once, the same for cracked wheat.

An electric mill would be easier, and so would a bread machine. There is nothing wrong with either—except that I prefer to have more of a hand in the work. In kneading bread, for example, I would miss the warmth and velvety softness of the dough, its scent, and the muscular rhythm of pressing, folding over, pressing …

Whole-wheat and Amaranth Bread

6½ cups whole-wheat flour
1½ cups amaranth flour
1 teaspoon salt
2 tablespoons honey
2 tablespoons dry yeast
2 tablespoons butter (or oil)
2½ cups water
Sesame seeds and egg white

Dissolve yeast in 3 ounces of the tepid water and I teaspoon of the honey. Cover and set in warm place until frothy (about 10 minutes). Dissolve salt and rest of honey in the remaining water. Meanwhile, rub butter into flour mixture. Add yeast mixture to the flour, then add the water—and mix to a dough. Knead on a floured surface until dough is no longer sticky (5-10 minutes). Place dough in an oiled bowl, cover with moist cloth, and set in a warm place until doubled in size (about 45 minutes). Punch down and knead until smooth (3 minutes). Divide dough into two oiled loaf pans. Brush tops with egg white and sprinkle generously with sesame seeds. Leave to prove until dough has risen (40 minutes). Bake at 425° F for 35-40 minutes, until well-risen and brown.

In a short time bread making becomes a habit, the steps requiring little attention. Heated briefly and then turned off, an oven makes a good "warm place," and so is a pool of sunlight. I am in the habit of setting the smaller bowl containing the yeast mixture in the larger bowl of liquid ingredients, and covering both. The no-knead breads are easier, and I relied on them in the beginning. Now, even though I am busier, this whole-wheat and amaranth is my daily bread, often with garlic and olive oil.

Garlic Olive Oil On Bread

Fill a cruet with olive oil, add a diced garlic clove. Will keep fine on the table. Excellent for whole-wheat bread instead of butter.

Cracker Bread

3½ cups whole-wheat flour
1½ cups white flour
1 cup amaranth flour
1 tablespoon dry yeast
1 teaspoon salt
1⅞ cups warm water
⅓ cup melted butter (or oil)
sesame seeds

Combine 4 cups flour, yeast, and salt, stirring well. Gradually add water, stirring well. Add butter, blending well. Gradually stir in enough flour to make a stiff dough. Knead until smooth and elastic (about 4 minutes). Place in a greased bowl and let rise until doubled in bulk (about 1 hour). Punch down and form dough into 9 balls, let rest 10 minutes. With a rolling pin, flatten each ball into a 10 inch circle, sprinkle liberally with sesame seeds and roll again. Place on a baking sheet. (Refrigerate extra dough balls.) Prick entire surface with a fork. Do not allow flatbread to rise. Bake at 325° F for 25 minutes—until lightly browned and crisp. (Optional: add cinnamon sugar or seasoned salt or grated cheese on top after rolled out, brushing surface lightly with water.) —Cracker bread is especially good with winter salads.

Amaranth-Blue Corn Dollars

¾ cup cornmeal
¼ cup amaranth flour
Less than ¼ teaspoon salt
¼ teaspoon honey

Measure ingredients into a bowl. Gradually add a little boiling water, stirring constantly until it is the consistency of waffle batter. Drop by spoon onto a well oiled, hot skillet. —Good with bean soup, buttered or plain. With honey or syrup, corn dollars are as good as any pancake. I usually keep some on hand in the refrigerator to reheat on the spur of the moment.

Hot Cereal

4 parts whole-wheat, coarsely ground

3 parts blue cornmeal
2 parts amaranth flour

One serving: ½+ cup cereal in 1+ cup water. Mix well in a saucepan and cook over low to moderate heat (10-15 minutes), stirring frequently. (I use a heavy iron saucepan, leaving the cereal to simmer over low heat—while I milk Isabel. On my return it's ready to serve after a final stir.)

Sesame Bars

2 cups sesame seed
⅓ cup honey
1/16 teaspoon salt

Mix together ingredients and spread in a ½ inch layer on a well-oiled 8 by 10 inch foil sheet (or pan with sides). Bake at 350° F for 35-40 minutes. While still warm cut into rectangular bars, let cool. —My staple snack food.

When I've been out working long out of doors and think of food —it is always of Rupert Brooke's "strong crust of friendly bread." And none is as friendly as that milled from the wheat I sowed in October and harvested in June.

DOING FOR ONESELF

The best thing about my work is that it is of my own choosing and done in my own way. Under those circumstances even the most menial work is pleasure.

Since I didn't know much about farming in the beginning, almost nothing, I made mistakes. But they were my mistakes to put right or live with. The work was difficult enough that I had to be inventive, but not so difficult I couldn't learn.

You cannot say to a crew of workmen, "Build me a farm." There is no such crew and no money to pay them. Just as there is no one else to do the work of the farm, no one else to make the decisions. The nearest equivalent to the small farmer is the housewife, especially if she is the mother of young children. We are amateurs, working for the pleasure of it rather than for hire.

There is some risk in doing for oneself. In taking on so much of our own work we sometimes take on too much, almost more than we can handle. But even that is better than taking on too little. The more of our affairs we hire others to do, the less confident we are about doing the rest—and so begins a circle of increasing dependency. As much as possible I want to keep my life simple enough that I can manage it.

I like to get on with a job so I can finish it—and spend some of each day at leisure, sometimes whole days. When we're not so busy we don't feel so important, and that's healthy too.

It's never a problem, these free hours. Sometimes I'm glad enough just to take a break from work, at the end of a row of beans lying flat on my back for awhile, watching the clouds drift overhead. In thoughtful idleness I've enjoyed some of the pleasantest moments of all.

In the country you prefer your own diversions. People assume I have a television. On long winter evenings I usually read, mostly older books—classic travelogues, natural history, and vintage adventure. Wilfred Thesiger, Laurens van der Post, John Muir, Henri Fabre, and John Buchan have had a great influence on my thinking. At meals or when working indoors I sometimes listen to short-wave, Radio Australia if I'm too late for the BBC newscast, the faint crackle of static giving world events an almost historical quality of distance and detachment. In the evening I enjoy Radio Netherlands. Most interesting by far, though, is the life of the field and the work of the farm.

As for boredom, the word has no meaning. It's inconceivable with so much to do and such an intriguing world to do it in. A small farm is part of the natural order, my work linked with that of the jumping spider I noticed on a hive cover in January. If the spider is active it means the bees will be too, scouting for winter dandelions and cleaning accumulated capping fragments from their hives—time for me to be checking the hives for winter losses due to mites, a cousin of that jumping spider.

From my kitchen window . . .

From my kitchen window I can see the garden and orchard, milk barn, hen house, beehives, and root cellar. Beyond the opposite wall is my pantry, and across the room is my dining table. On a small farm, one that aspires to a reasonable self-sufficiency, the kitchen is the center of activity. On my infrequent absences from the place I have sometimes found myself at a loss. But as soon as I was in a kitchen I was on familiar ground. Here I could make myself useful. I knew what to do.

As many as seven months have gone by without my going to a grocery store. Except for the obvious—items like spices, baking soda, olive oil—everything comes from the farm. I cook with whatever is on hand, either fresh or preserved, improvising when I don't have all the ingredients. For the most part my menus are seasonal—in spring, creamed asparagus on toast, new potatoes and peas; in summer, steamed sweet corn and okra (sliced thin, rolled in cornmeal, and fried); in fall and winter, soups and baked chicken and winter squash patties (with butter, honey, and spices). But I also have my perennial favorites—

Breakfast: hot cereal, a glass of cultured milk, a poached egg on a bed of popcorn (adding a little vinegar to the egg's cooking water to reduce froth), and fruit (cantaloupe, peaches, apples, frozen blackberries, fresh mulberries and cream, depending on the season). Sometimes, instead of hot cereal, I have a big bowl of popcorn with honey and plenty of milk. At a respectful distance from my plate one of the cats watches solemnly. And I listen to the morning newscast on BBC.

Lunch: a cheese sandwich with lettuce or sprouts or sliced cucumber and a bowl of soup (zucchini, tomato, most often corn chowder). Either that or a bowl of bean soup with amaranth corn dollars. Usually there are leftovers from the evening before, like fried okra. And always there is bread, to dip in olive oil and garlic or spread with butter and honeycomb, and fruit. By lunch time I will have put in half a day's work, so I linger over a book propped on the table.

Dinner: salad, a casserole or chicken (a half-cup serving rolled in whole wheat flour and fried or curried eggs or curried chicken on toast), mashed potatoes or baked winter squash, and a vegetable ... from late June into September, fresh steamed corn, stuffed zucchini, fried okra. Always there is bread and fruit, like applesauce or frozen berries. And with my book and a cup of tea and Olive curled up in my lap, I am at the table at least an hour. Even when there is work to do in the evening, I take my time.

It would never occur to me I was wasting time. I like everything about the kitchen—the restaurant china on open shelves, the warm scent of cooking apples, the squeak of the rolling pin. On a small farm there is no real distinction between kinds of work, even between work and leisure. I enjoy housekeeping. Even the laundry, which needs doing frequently on the farm, gives me a sense of accomplishment and, in the spectacle of clothes on the clothesline snapping and billowing in a March breeze, something approaching ecstasy.

But meals come around three times a day, and I have wondered why cooking isn't tedious. Food would be tedious if it really came in boxes, but it doesn't. The chemistry of recipes fascinates me, and I sometimes read labels on boxes in the store. The ingredients are too many, most of them in no kitchen. And there isn't the pleasure of knowing the whole history—from the time it was a seed and a stream of milk sizzling against the side of the pail until, on a white china plate ringed with green lines, it is served up as creamed eggplant.

Creamed Eggplant

1 large eggplant (or 2 medium)
3 tablespoons butter
⅓ cup flour
1 cup milk (or cream)
⅓ cup grated cheese
¼ teaspoon salt
pepper to taste
2 teaspoons oregano

While the whole, unpeeled eggplant is baking in a 350° F oven until tender (45 minutes to 1 hour), prepare the sauce. Melt the butter, stir in the flour until smooth, slowly add the milk. When smooth, melt the cheese in the sauce and add the salt, pepper, and oregano. Skin and mash the eggplant, mix with the sauce, reheating if necessary. Serve hot.

Eggplant, with its exotic purple or ivory fruits, is easy to grow and I enjoy having it in the garden. But finding a place for it on the table wasn't as easy, until I learned about creamed eggplant. Visitors at mealtime are surprised I go to so much trouble when there is just myself. But it's partly because of dishes like creamed eggplant that I enjoy the kitchen so much. Even green tomato pie is worth the extra time.

Green Tomato Pie

¾ cup honey
½ teaspoon cloves
½ teaspoon allspice (or nutmeg)
1 teaspoon cinnamon
⅛ teaspoon salt
¼ teaspoon vinegar
½ cup cornstarch (or flour)
4 cups green tomatoes, cored and cut into 8ths or 16ths

Combine ingredients in a 2-quart saucepan, stirring over low heat until well mixed. Brush bottom of prepared unbaked pie shell with egg white. Pour warm pie filling into the shell and smooth, adding a well pricked top crust. Bake at 325°-350° F for 55-60 minutes.

Sweet, Whole-Grain Pie Dough

1 cup flour
¾ cup whole-wheat flour
¼ cup amaranth flour
¼ cup sugar
½ teaspoon salt
¾ cup softened butter

2 egg yolks
1 teaspoon vanilla

Combine flours, sugar, salt. Using the fingers, work in softened butter. Add egg yolks and vanilla, stirring with fingers until mixture forms a ball and loses its stickiness. Cover and chill in refrigerator for at least 30 minutes. Bring to room temperature before rolling half the dough for a bottom crust, placing this in a greased pie pan. Roll the remaining for a top crust.

Despite the green tomato pie, I think of myself as eating simply—and what the farm produces. Cooked when fresh from the garden, turnips are sweeter and more tender than their reputation. Left in the ground, turnips will keep well into December, and clamped they will keep through the winter. (To clamp: place turnips on a bed of hay in a dry hole and cover them over with a low mound of hay, a layer of soil on top.) Hot turnip slaw, an old family recipe from Germany, is a bracing cold weather dish.

Hot Turnip Slaw

4½ quarts sliced turnips (thinly sliced in grater)
½ to ⅔ cup vinegar
1 teaspoon salt

¼ (or more) teaspoon pepper
Stock of well-cooked giblets or roast chicken
(Total liquid should equal 4-5 cups)

Combine stock, vinegar, and seasonings. Add the thinly sliced turnips and cook slowly in a covered pot until well done but not mushy, as much as 3 hours if cooked very slowly. Makes 12 to 15 servings.

Marinated cucumber is an easy salad. I keep a large, covered bowl in the refrigerator all summer long—replenishing whatever seems to be running short until time for a fresh start.

Marinated Cucumbers

4 sliced cucumbers, picked before they are too large
1 sliced onion
1 sliced medium green pepper
½ cup vinegar
⅜ cup honey
1 quart water

Combine vinegar, honey, and water. Add cucumbers, onion, and pepper. Store in refrigerator. Adding an ice cube to each serving will keep it cold at the table.

Baked winter squash is even easier. Stored in a cool place with half an inch or so of the stem left as a seal, butternut squash (my favorite) will easily keep until summer, when the next crop is ready to pick.

Baked Winter Squash

Using a cleaver or other heavy knife, cut the squash in half lengthwise, and remove the seeds. (A sharp rap with the dull edge of the knife blade will first knock away the dried stem.) Place cut side down in a baking dish, add ½ inch water. Bake at 350° F for 45 minutes, or until tender. Served "on the half shell" it is mealy and sweet—more so with butter, spices, and honey. I prefer winter squash pie to pumpkin.

Winter Squash Pie

Filling for 2 pies:
4 cups cooked winter squash (butternut is best)
2½ cups half-and-half (cream and milk)
1¼ cups honey
2 teaspoons cinnamon
1 teaspoon ginger
½ teaspoon nutmeg
¼ teaspoon cloves
4 beaten eggs

Refrigerated it will keep for days and frozen in

serving size pieces even longer. Virtually a light meal in itself, squash pie is one of my year round staples.

Other staples are chowders and soups. Along with cheese, bread or corn dollars, and fruit—I can usually put together a substantial meal in just a few minutes from what is on hand. Recipes for both corn chowder and bean soups are in most cookbooks, but not for zucchini soup.

Zucchini Soup

- 2 large onions or leeks, minced
- 2 tablespoons butter
- 2 potatoes, peeled and sliced
- 4 cups chicken stock
- 1½ tablespoons vinegar
- 1-2 garlic cloves, minced
- ⅛ teaspoon pepper
- 2 pounds zucchini, sliced thinly (or 2 quarts canned zucchini, drained)

In a kettle, cook onions in butter over moderate heat for 5 minutes or until softened. Add potatoes, chicken stock, vinegar, garlic clove, and seasonings. Bring liquid to a boil, stirring occasionally. Add zucchini and simmer the soup, covered, for 30 minutes. Serve hot. Good with hard rolls.

On every table in the dining room of the Kenricia Hotel (in Kenora, Ontario) there used to be a basket heaped with hard rolls. The hotel was an old frame building, and tied to the radiator in every second storey room next to the window was a coiled rope the thickness of my arm—the fire escape. And somewhere there were always voices singing "Oh, Canada." But most fascinating for a child were those hard rolls, round as cannon balls and unforgettably good.

Kenricia Hard Rolls
(whole grain version)

4¼ cups flour

¾ cup amaranth flour

½ cup whole-wheat flour

¾ teaspoon salt

½ cup lukewarm water

1½ tablespoons honey

2 packages dry yeast

⅞ cup milk

1 egg

2 tablespoons butter

Dissolve yeast in lukewarm water, add honey and let froth. Sift together flours and salt. Beat egg and add to lukewarm water. Add liquids to flour, mix to a dough and work in softened butter. Make a stiff dough and knead until smooth (10 minutes). Leave to rise in a greased, covered bowl until doubled in bulk (less than 1 hour). Punch down and knead until smooth again (3 minutes). Take small pieces of dough and place in ungreased muffin tins—makes about 24 rolls. Allow to prove (rise) 30 minutes. Bake at 450° F for about 25 minutes, until well-browned.

Since corn and beans together are a complete protein, cornbean pie is a natural—and one of my staple dishes.

Cornbean Pie

Crust:
2 cups cornmeal
3 tablespoons oil
½ cup hot stock

Filling:
1 medium onion, chopped
½ cup Swiss chard stalks (or sprouts)
1 minced garlic clove
½ cup chopped green peppers
2 tablespoons oil
2½ cups cooked beans
pinch of cayenne

pinch of celery seed
2 teaspoons cumin
½ cup chopped tomatoes
2 tablespoons soy sauce
⅓ cup grated cheese

Mix crust ingredients and pat into an oiled 9 inch pie pan. Sauté onion, Swiss chard, garlic, and green peppers about 5 minutes. Add beans and spices, warm the mixture, and pat into cornmeal crust. Combine tomatoes and soy sauce, and pour over the bean mixture. Bake 25 minutes at 350° F. Sprinkle with the cheese and bake 5 minutes or longer.

A winter salad, which goes well with Cornbean Pie, I make from sprouts and whatever is growing in the cold frames. It's a form of year-round gardening with never a crop failure—and with cheese a perfect meal.

Winter Salad

Assorted greens, cut up (lettuce, spinach, kale)
Sprouts (mung bean and/or canola)
Cubed cheese, if desired
Oil and vinegar dressing
Sprinkle with bread crumbs, popcorn, or toasted sesame seeds

Sprouts: Pour ⅓ cup mung beans or canola seed into a wide-mouth quart jar, fill jar with lukewarm water and let stand overnight; use a screen wire lid for canola, ⅛″ galvanized hardware cloth for mung beans. Drain in the morning and rinse; rinse twice daily until seeds sprout enough to fill the jar; drain for several minutes after rinsing. Between rinsings, roll jar in a towel, tip bottom end of jar up slightly for additional draining. When sprouts are ready, replace lid with regular cap and store in refrigerator.

Necessity has made a cook of me—necessity and the abundance from the farm. And as with everything else, after I got into it I discovered there was a great deal of fun to be had. Once I had occasion to enjoy some Italian dipping cookies which were said on the package to have originated in 13th Century Tuscany. With the list of ingredients I worked out the recipe.

Dipping Cookies
(Biscotti Toscani)

3 cups flour
3 teaspoons baking soda
½ cup sugar
4 eggs
½ cup milk

2 teaspoons almond extract
⅔ cup whole raw almonds

Beat eggs, add sugar then milk and almond extract. Sift and measure flour, resift with baking soda. Combine egg mixture with dry ingredients and mix well, stir in almonds. Pour into 2 greased 9x5 inch loaf pans and bake at 325° F for 40 to 45 minutes. Let cool, remove from bread pans and cool further. Using a bread knife, slice ½ inch thick. Space slices on wire rack and bake at 325° F for 35 minutes, until crisp. Keeps well.

Cooking with Honey

In cooking, I ordinarily prefer honey to sugar. Since honey is already a liquid, it is often easier to use than sugar—and adds a subtlety of its own. In cookies (and in some cakes) it is best to replace just one half the sugar with honey, otherwise they may be soggy and hard to get off the cookie sheet. But in most cases honey can easily replace sugar.

Three-quarter cup honey equals I cup sugar. Reduce liquids ¼ cup for every cup of honey used. If honey crystallizes, place jar in 130° F water. In baked foods add ½ teaspoon baking soda for every cup honey used. Lower baking temperature 25° F and bake slightly longer.

Much of the work of a small farm is carried on in the kitchen. This is where the pail of milk comes from the barn, the basket of eggs and dressed poultry, and all the produce from the garden and orchard. Some of the milk I culture, producing a thick drink much like yogurt.

Cultured Milk

Fill a quart jar with milk (skimmed or homogenized). Cover with water and bring to a boil. Boil for 30 minutes — pasteurizing to keep the culture pure. Cool milk to room temperature. Add 2-3 tablespoons from the previous batch (or of live yogurt culture) and shake well. Roll up in a towel and lay on its side in a draft-free place to incubate overnight. Store in refrigerator.

Milk freezes well if the cream is first skimmed off and plenty of head space left in the container. The cream, and there is a high percentage in Jersey milk, I churn into butter. Butter made from Jersey cream, which contains more carotene, is bright yellow.

Cheese is more involved, at least there are more steps in the process. But with a little attention now and then while I'm engaged in some other work around the house, I can make a fine cheese in a day—either "Farmhouse" or "Swiss."

Since I put up quite a number of cheeses, I always write the date each was made in the soft wax coating. There is always an element of expectation when I cut into a cheese for the first time. While fairly uniform, the longer a cheese has aged the firmer its texture and stronger its flavor. Nothing I do in the kitchen gives me quite the satisfaction as those mild and creamy cheeses.

Easiest to preserve for future use are foods I can simply refrigerate—like beets, garlic, radishes, apples, and those aging cheeses. Milk, butter, pecans, some berries, watermelon juice, and sliced green peppers are easy to freeze. Most other foods take some preparation.

Eggs I warm to room temperature, blend (adding 1 teaspoon honey for each 5 eggs as an emulsifier to keep them from separating), and freeze in ice cube trays. Since 1½ cubes equals 1 egg, I cut some of the cubes in half before storing loosely in plastic bags. Dressed hens I freeze whole, filling the bag with water to prevent freezer burn.

Canning is something of a lost art, though I don't know why. Some fruits and vegetables seem to benefit from having been canned. I simply sort, wash, and trim vegetables from the garden or fruits from the orchard as if getting ready to cook them for dinner. Instead, I heat the vegetables, fill the canning jars, cover with boiling water, and place in a canner (where air is exhausted and the heat rises to 250° F at 15 pounds pressure, 5 pounds for fruit because of its higher acid content). This heat deactivates enzymes and kills molds, yeast, and bacteria. As the jars begin cooling after removal from the canner, the lids self-seal. Gleaming on the pantry shelves, their contents clear and vividly colorful, the jars of fruits and vegetables are quite decorative.

Visitors in the fall, at the end of canning season, inevitably ask to see the pantry. And I enjoy the sight of all that plenty myself. There are also jars of herbs—first gathered in the morning while essential oils are at their peak and hung to dry in the pantry.

All the various dry beans and peas I harvest, dry and store as they are, after first heating in a pan. Twenty minutes in the oven at 150° F kills any insect eggs they harbor.

Garlic, okra, pears, and apples I slice thinly and spread on screen frames—outdoors to dry in the sun. The garlic I then grind in a blender to make a strong garlic salt (60% dried garlic, 40% salt). Okra sun dries in three or four days, retaining its fragrance and clear green color, that of well-cured hay.

A staple is fruit leather, apple and peach. The process is quite simple: cook until soft, pulp in a blender, add spices and honey, spread on oiled freezer paper taped to plywood, and dry in the sun. Heat builds up so quickly flies and bees are not a problem, and it's brought indoors at night. In a couple of days the dried fruit leather should be ready to roll up and bag. It seems to keep indefinitely in my pantry, which is reasonably cool.

The bright, hot days of high summer with their light southerly breezes are ideal for drying. And they come just at the end of harvest.

Honey stores best at room temperature, and in October I move the remaining cases indoors. Until sales reduce my inventory, nearly a ton of it is stored in the closets and living room. Covered with an old Hudson Bay blanket, the labeled quarts of honey are a mountain for the cats to play on—until Christmas when it's mostly sold.

Onions, potatoes, and winter squash are stored in the root cellar, the potatoes in moist sand. Pears have never been long keepers for me, coming on when it's still warm, even in the cellar. But apples, especially the Arkansas black, keep well nearly all winter.

It is not out of any vague fear of scarcity—mine not a survivalist's mentality—but I do take satisfaction in a full larder and in a bright, warm kitchen and in sitting down to a meal of my own from start to finish.

F R U G A L I T Y

With so much to recommend a small farm there must be a catch, and there is. Any work that brings in so little income is not good business. Labor intensive as well and requiring an initial investment, it is a very bad business.

Fifteen dollars earned hatching a neighbor's peacock eggs under a broody hen doesn't count for much in a cash economy, less in a society where income is a way of keeping score. If I had any illusions about money—or about anything else, the farm a good place to learn to face unpleasant facts—I quickly got over them.

The farm carried a price tag. With money I had saved, I was able to put the farm together without going into debt—this, by building a house and barn for little more than some of my school friends spent for a car. Still, it was with this advantage I began, good luck for which I'm deeply grateful. It's stimulating living hand to mouth, season to season—but not if burdened with debt.

Running a small farm is an equation that calls for careful management, but one I consider worth the risk.—Security, comfort, convenience are to be valued only to the extent they do not deprive us of something better.

I learned to advantage that it's better to own fewer things—of higher quality. My English spading fork, hand forged and costing two-and-a-half times that of a mass produced one, will last a lifetime and is such a thing of beauty I never clean the tines, burnished from use to a silvery sheen, without thinking how fine it would look on my wall.

I make a point of going to the most basic level I can in the production chain. A 50 lb. sack of clean, fresh wheat costs around $4.00 at the local co-op and a 2 lb. bag of yeast $3.20. For less than 20 cents I can bake a 2 lb. loaf of bread.

Passing up the discount outlet, I prefer the small tradesman—where I can get good advice about what I need for a job around the farm and how to go about it.

But the most important lesson I learned is—I'm better off doing with less. What order and serenity

I've found comes not so much from living on a small farm as from the practice of frugality.

That the life of my choice . . .